Agroforestry Systems for Resource Conservation and Livelihood Security in Lower Himalayas

nipa books for everyone PAPERBACKS

NIPA® GENX ELECTRONIC RESOURCES & SOLUTIONS P. LTD.
New Delhi-110 034

www.nipaers.com
Browse, Search, Read & Buy...
Print Books
eChapters
eBooks
Publishers
Forthcoming
Subject Catalogues
New Titles
Online Resources on:
✓ Current Affairs
✓ Reasoning, Logic & Aptitude
✓ Competitive Examinations
✓ English Language Lab
✓ Writing & Pronunciation Tools
✓ Personality Development
✓ Online Programmes for Professionals
✓ Interview Preparation
Pay Using
PhonePe
Google Pay
Paytm
PayPal
UPI
CCAvenue

Agroforestry Systems for Resource Conservation and Livelihood Security in Lower Himalayas

Editors

PANKAJ PANWAR

A.K. TIWARI

K.S. DADHWAL

nipa® books for everyone PAPERBACKS

NIPA® GENX ELECTRONIC RESOURCES & SOLUTIONS P. LTD.

New Delhi-110 034

nipa books for everyone
PAPERBACKS

NIPA® GENX ELECTRONIC RESOURCES & SOLUTIONS P. LTD.

101,103, Vikas Surya Plaza, CU Block
L.S.C. Market, Pitam Pura, New Delhi-110 034
Ph : +91 11 27341616, 27341717, 27341718
E-mail: newindiapublishingagency@gmail.com
www: www.nipabooks.com

For customer assistance, please contact
Phone: + 91-11-27 34 17 17
Fax: + 91-11- 27 34 16 16
E-Mail: feedbacks@nipabooks.com

ISBN: 978-81-96079-09-3

Composed and Designed by NIPA®.

PREFACE

The Lower Himalayas also known as the Shivalik ranges cover a large portion of Himachal Pradesh, UttaraKhand, Jammu and Kashmir, Punjab and Haryana states of north-west India. The region covers an area of 2.14 million ha and forms strips, which stretches 530 km in length with width varying from 25 to 95 km. Once covered with lush green dense forest cover, most of the Shivalik hills today represent eroded ravines, barren slopes and ugly scenes. Every year enormous quantity of soil is eroded resulting in the rise of riverbeds and siltation of tanks, reservoirs and other natural water bodies. Only 18 per cent of the cultivated area in Shivaliks is irrigated. At present Shivaliks are the hot spot of erosion in the world. Soil erosion greater than 80 t $ha^{-1}yr^{-1}$ have been recorded from denuded hills at places. In one of the assessment for erosion from Himachal Pradesh, soil loss up to 682 t $ha^{-1}yr^{-1}$ has been estimated in the lower Himalayas. Hence the need is to arrest soil, water and nutrient losses for resource conservation and sustainable production and for that agroforestry is one of the viable option for Shivaliks.

Agroforestry has become a widely used 'term of art' in agriculture and forest policy. It also has become a widely used political and public relation tool. National Agriculture Policy 2000 of India has emphasized the need for diversification in agriculture through agroforestry. Similarly, the Task Force on Greening India for Livelihood Security and Sustainable Development, cover Joint Forest Management and Agroforestry as the viable options. Today agroforestry has become an important land use strategy and has emerged as one of the sustainable option of land management.

Research institutes, universities and other research organizations working in the region are strongly recommending agroforestry as a viable option for such undulating, rainfed and erosion prone areas. Taking into account the importance of agroforestry in livelihood security and resource conservation in Shivaliks, it was thought to club the work done in different organizations on the subject so that a ready reference of literature for researchers, planners and other inquisitive reader is available.

We feel that the authors of the chapters have done justice in compiling, collating and synthesizing the research and technologies developed in their respective regions and area of specialization. The book will certainly benefit the researchers, progressive farmers, policy makers and other development agencies. We appreciate the commitment of the contributing authors.

We wish to thank all the contributors for complying with the time schedule and bearing with the persistent requests and queries made by us. We would like to extend our deep appreciation to former Director Dr. V.N. Sharda, Central Soil and Water Conservation Research and Training Institute, Dehra Dun for his continuous guidance and support.

Pankaj Panwar
A.K. Tiwari
K.S. Dadhwal

CONTENTS

Chapter-1

Agroforestry Scenario and Opportunities in Shivalik Himalayas

PANKAJ PANWAR, A.K. TIWARI AND K.S. DADHWAL*
Central Soil and Water Conservation Research and Training Institute
Research Centre, Chandigarh- 16 0019
* Central Soil and Water Conservation Research and Training Institute,
Dehradun-248195

ABSTRACT

More than 80 per cent area of the Shivaliks is rainfed. Soil erosion greater than 80 t $ha^{-1}yr^{-1}$ is being recorded from denuded hills. Though, trees are nurtured on the farms but no definite pattern of trees on field bunds is followed, (neither from spacing point of view nor from species point of view). Further, these trees are not silviculturally managed to get best output from tree components as well as for posing less competition to the crops. Shivaliks has a wide range of climate, geology, soil, topography and socio-economic strata coupled with wide variety of vegetation. Such variations provide ample opportunities for developing site specific agroforestry systems.

INTRODUCTION

The Lower Himalayas also known as the Shivalik region ranges from 30^0 10′ and $33^037'$ N latitudes to $73^037'$ and $77^039'$ E longitudes and cover a large portion of Himachal Pradesh, Uttarakhand, Jammu and Kashmir, Punjab and Haryana states of north-west India. The region covers an area of 2.14 m ha and forms strips, which stretches

530 km in length with width varying from 25 to 95 km (Kukal *et al.*, 1991) (Fig. 1). Prior to the middle of 18th century, the Shivalik hills were strictly preserved for hunting and no cultivation, grazing or exploitation of timber was permitted. At that time the hills were covered with thick lush acacia, dalbergia (shisham) and pine forests with a profuse undercover of shrubs and grasses along with wildlife. However, after the Sikh wars (1845–49) the Sardars and Rajas, who owned the hunting lands, were evicted and the forests were handed over to the villagers in addition to their village common lands. In less than two generations, unrestricted tree felling and overgrazing have played havoc with the vegetation. Despite of Punjab Land Preservation (*Chos*) Act, 1900 which restricts grazing and prohibits tree felling, the natural resources of Shivaliks are in the process of severe degradation and deforestation basically due to increased biotic pressure that has its roots in the complex web of social, economic and institutional problems. Most of the Shivalik hills today, represent eroded ravines and barren slopes. Every year enormous quantity of soil is eroded resulting in the rise of riverbeds and siltation of tanks, reservoirs and other natural water bodies. Only 18 per cent of the cultivated area in Shivaliks is irrigated. Undulating topography, cultivation on steep slopes, age old agricultural practices and frequent crop failures, due to rainfed condition, force the people to keep large herd of cattle. However, the availability of fodder is only one third of its requirement. Grasslands are infested with abnoxious weeds like *Lantana camara* and *Parthenium* spp. which have eliminated both fodder as well as commercial grasses. Exploitation of natural resources and vegetation, denudation of catchments, drying up of perennial springs and stream and ground water depletion are the associated problems cropping up due to deforestation and high biotic pressure in the Shivalik region. As a result, of 94.63 lakh ha degraded land, 26.94 lakh ha (assessed during 2003) was degraded due to soil erosion caused by water. The eroded soil when settles in area down stream renders them unfit for cultivation and also reduce the water storing capacity of water harvesting structure. Arya and Samra, 2001 had reported that around 38 per cent of water harvesting structures failed in Haryana immediately after implementation, due to siltation of dam and chocking of pipeline.

At present Shivaliks are the hot spot of erosion in the world. Soil erosion greater than 80 t $ha^{-1}yr^{-1}$ is being recorded from denuded hills of Shivaliks. The assessment of erosion for Himachal Pradesh estimated soil loss up to 682 t ha^{-1}yr^{-1} in the lower Himalayas (Tiwari *et al.*, 2005). Such high rates of soil erosion are rendering the lower Himalayas unfit for cultivation.

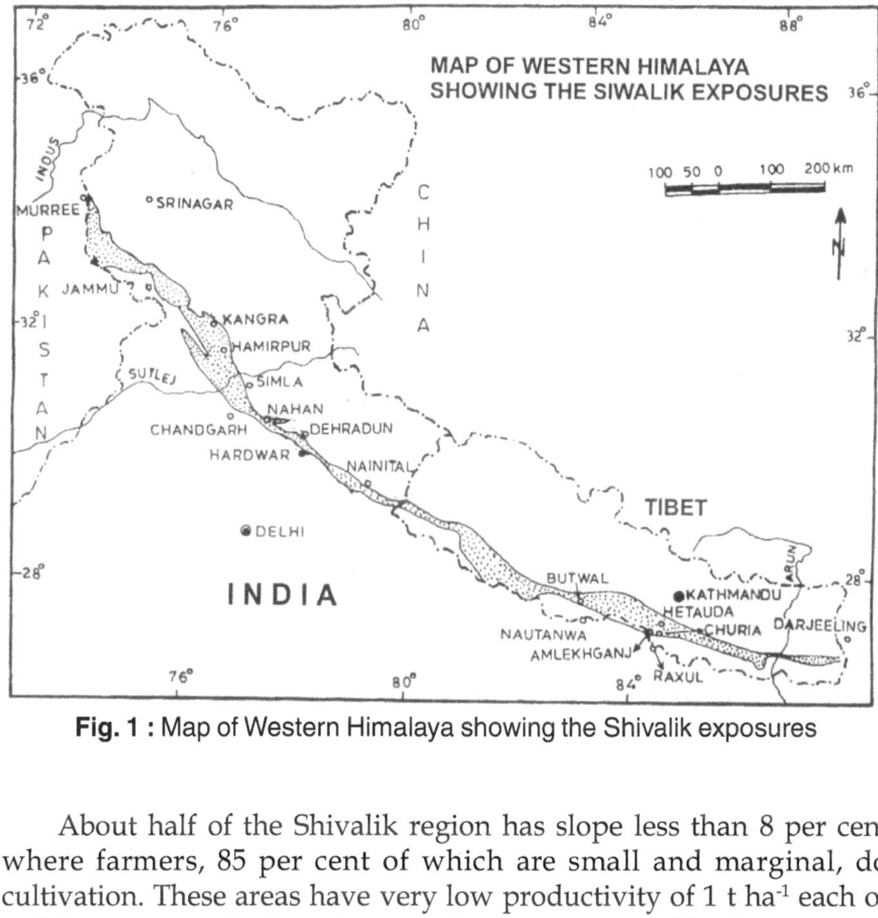

Fig. 1 : Map of Western Himalaya showing the Shivalik exposures

About half of the Shivalik region has slope less than 8 per cent where farmers, 85 per cent of which are small and marginal, do cultivation. These areas have very low productivity of 1 t ha^{-1} each of paddy, maize and wheat in a good monsoon year. The late onset of monsoon, its early withdrawal and stress during *kharif* cause marked reduction in crop production even in 20 per cent of the good monsoon years.

Traditionally, trees and crops were grown in association with each other in Shivaliks. The reason was to get multiple output from same piece of land, though such systems were exclusively meant for sustenance with no commercial intent. As the population increased there was a need to produce more food and hence slowly the tree component vanished from the farmlands and monocropping became dominant form of agriculture system. The need of the hour is to accept the trees as an integral part of the farming system in Shivaliks to revert the soil degradation and erosion and also to reduce the dependence of external inputs in production system. ICRAF, 2006 has clearly indicated that "Tree based farming system / agroforestry is a dynamic, ecologically based, natural resources management system that, through the integration of trees on farms and in the agricultural landscape, diversifies and sustains production for increased social, economic and environmental benefits for land users at all level".

Natural Resources of Shivaliks : Climate, Geology & Soils and Water Resources

Climate

The region receives average rainfall of 1100 mm yet length of growing period is 100-150 days depending upon rainfall quantity and distribution, soil, topography and micro-climate. Even in a good rainfall year, it can support only one good crop during *kharif.* However, the rainfall distribution is too erratic within and among years. Rainfall trend at Panchkula (Haryana) for 42 years (1958-99) revealed that it received average rainfall of 1124 mm and ranged from 592 to 1892 mm during the period (Agnihotri, 2000). Out of 42 years, the area experienced 8 large and 5 severe droughts. Thus one out of three years had severe rainfall deficit which adversely affected the crops during *kharif.* The temperature has a wide range of 5.6 ^{0}C to 31 ^{0}C in winter and 15.8 ^{0}C to 40.4 ^{0}C in summer on weekly average basis.

Geology and soils

The Shivalik region comprises of four major physiographic zones *viz.*, hill top/slope, piedmonts, alluvial plains and valleys having slope >15, 8-15, 3-8 and 1-3 per cent, respectively. On the basis of lithology and fossil contents, Shivalik region has been broadly divided into three divisions *i.e.* lower, middle and upper Shivaliks ranging in age from Middle Miocene to Upper Pliocene. Lower Shivaliks are characterized by grey to light grey, micaceous fine to medium sand stone occasionally with psuedo conglomerates containing pebbles of calcareous clay and shale. Middle Shivaliks comprise of reddish brown, grey medium grained soft sand stone interbedded with red, orange and yellow clays and sand stones occasionally with psuedo conglomerates. Upper Shivaliks are represented by Pinjore sand rocks, boulder conglomerates interbedded with sand stone and clays.

Soils on hill tops belong to loamy-skeletal, Typic Udorthents and Lithic Udorthents families. Soils on hill sides/reposed slopes are loamy, Typic Udorthents and Dystric Eutrochrepts. Terraced lands have more developed loamy, Typic Eutrochrepts. Piedmonts have relatively young soils and are classified as sandy, Typic Ustorthents and sandy-skeletal/ loamy-skeletal/loamy, Typic Ustorthents. Valleys are dominated by loamy-over-fragmental, Typic Ustifluvents. Distribution of soil families in entire Shivalik have also been mapped (Sidhu *et al.*, 2000)

Majority of the soils (64%) are moderately deep (50-75 cm) and 19 per cent are deep having depth more than 75 cm. Only 15 per cent area has soils with depth less than 50 cm. Texture wise, two-third soils are loamy in nature without gravel and 28 per cent with loamy texture having more than 25 per cent gravels. Only 1.7 per cent soils have sandy texture and none are clayey in nature. The soils have wide range in organic carbon content (0.1 to 1.1%) and cation exchange capacity (3 to 20 cmol (p^+) kg^{-1}). While the soils have the potentials for large varieties of crop and plant species these require ameliorative soil conservation measures to enhance their production potentials and sustain at the optimum level (Yadav *et al.,* 2004).

Water resources

Shivalik region receives rainfall in the range of 800 to 2500 mm annually. The amount of rainfall generally increases with elevation. Average rainfall for the region works out to be 1100 mm and majority of it belong to dry, sub-humid region (annual rainfall 1100-1600mm) with some lower Shivalik region being under wet, semi-arid and some upper Shivaliks in Himachal Pradesh under wet, sub-humid. The dry months with monthly rainfall less than 30 mm varies between 3 and 5 months in a year giving rise to ustic and udic moisture soil regimes.

High intensity rainfall with rolling topography induces large volume of runoff giving rise to various seasonal and perennial rivers. Shivalik region is endowed with vast and sustained water resources and the area serves as catchment to many big rivers (Yamuna, Beas, Ravi, Satluj and Chenab), large reservoirs (Bhakra, Pong, Thein), lakes (Renuka, Sukhna) and small dams and ponds. Proper management of water resources besides meeting local needs with respect to drinking water, irrigation, electricity generation etc. can be and is being spared for other regions of the country.

Agroforestry Scenario in Shivaliks

Keeping fodder, fuel, timber and fibre yielding trees on farms of Shivaliks is a common practice and way of life. However, no definite pattern of tree nurturing on field bunds is followed, neither from spacing point of view nor from species point of view. The bunds of terraces may have only two to three trees or it may have large number of trees (Photo 1). The species and number of trees retained on the field bunds invariably depends upon the number of cattle, distance from the house and availability of community and/or forest land adjoining the villages. As a thumb rule, more the milch animals more number of fodder yielding trees are retained on the farm as the tree fodder is required during winter when grasses are scarce. The agricultural fields which are far from the house tend to have more number of trees as it is difficult to keep continuous vigil from wild animals damage if crops are to be grown. Hence, trees are the best choice for such unattended areas and also requires less or to say no input at all. Availability of community lands and/or forest lands adjoining the village also affects the presence of trees on the farm. If community land and/or forest are nearby, the requirement of fodder, fuel, timber and other forest/ tree based needs are met out from these

Photo 1 : Irregular pattern of tree being nurtured on the farms in Shivaliks

common lands and less number of trees are maintained on the farms. Thus the presence of trees on field invariably depends upon farmer's compulsion and/or opportunity then his wisdom of trees benefits on soil, water and environmental conservation. Rather, farmers perceive that trees are competitive in nature and cannot have supplementary or complementary effect.

Despite all the opposition for not inclusion of trees on the farmlands many of the trees are still nurtured on farm. The list of such species is presented in Table 1.

Table 1 : Trees nurtured on the field boundaries in Shivaliks

Scientific Name	Common name	Scientific Name	Common name
		A. Fodder yielding trees	
Middle and Upper Shivaliks		**Lower Shivaliks**	
Grewia optiva	Beul/Bhimal	*Grewia optiva*	Beul/Bhimal
Celtis australis	Khirak	*Morus laevigata*	tut
Bauhinia variegata	Kachnar	*Anogeissus latifolia*	Chalil
Morus alba	Shahtoot	*Bauhinia variegata*	kachnar
Prunus puddus	paja		
		B. Timber yielding trees	
Toona ciliata	Tuni/Toon	*Toona ciliata*	Tuni
Melia azedarach	derech	*Ulmus villosa*	Ulmus
Pistacia integerrima	kakkar	*Melia azedarach*	Derech
		Albizia lebbeck	Kala siris
		Albizia chinensis	Chuhi
		Acacia catechu	Khair
		Bombax ceiba	Semul
		C. Fruit yielding trees	
Prunus saliciana	Plum	*Mangifera indica*	Mango
Prunus amygadalus	Apricot	*Pseudium guvajava*	Guava
Prunus communis	Pear	*Litchi chinensis*	Litchi
Diospyros kaki	Persimmon	*Citrus lemon*	lemon
Prunus persica	Peach	*Terminalia bellirica* and *T. chebula*	Bhera and harad
Pyrus pashia	kainth	*Embilica officinalis*	Aonla

Need of the hour

Most smallholder agroforestry systems are characterized by limited proactive management and planning. Spacing is irregular and species components often primarily the result of chance. Harvesting products is often the most common management activity, with minimal weeding to control herbaceous and woody competition. As a result, the quality and quantity of products may be far below the systems' potential. The productivity of most smallholder agroforestry systems can be improved by enhancing smallholder management skills. Key skills include: species selection/site matching; identifying tree farming systems that match farmers' land, labor and socioeconomic limitations - including annual crops, tree crops, intercropping and understorey cropping options; tree management options to produce high quality products; pest and disease management; and soil management. Efforts should seek to develop a range of deliberate management techniques for trees and systems that enable farmers to produce quality products for specific market opportunities.

Agroforestry Opportunities in Shivaliks

Shivaliks has a wide range of climate, geology, soil and topography. The socio-economic strata also vary widely in the region. As a result the whole region has wide variety of vegetation. Due to these opportunities large population had settled in the region in the past century. The huge population pressure had resulted in large scale exploitation of the natural resources of the region thereby diminishing the resources to such an extent that the time had come when many of the species are on verge of extinction or at least became threatened. Associated problem is of excessive soil erosion.

To tap the opportunities available in the region number of institutes *viz.* Central Soil and Water Conservation Research and Training Institute, Chandigarh Centre; Y.S. Parmar University of Horticulture and forest, Solan; CSK Agricultural University, Palampur; CSWCRTI, Dehradun; Panjab Agricultural University-Regional centre, Balowal Sonkhari, SKUAST- Jammu; Soil conservation department and forestry departments of Himachal Pradesh, Haryana, Panjab, Uttarakhand, Jammu and Kashmir are working on developing different landuse system and practices involving mechanical and vegetative measures to improve the productivity and production of the region to fulfill the need of the population on one hand and on the other sustain the ecosystem. Different systems evolved and recommended by various agencies have been presented in the succeeding chapters of the book.

Shivaliks provide varied opportunities to work in agroforestry. These opportunities are discussed under the following subheads:

Diversity of species

Shivaliks have vide variation in altitude, aspect, slope and hence soil. As a result of these variations different types of species had evolved over a period of time in specific microclimate. Therefore, as such Shivaliks had varied choices of trees right from timber yielding, fruit plants, fodder, fuel and medicinal plants. The diversity of these species provide lot of opportunity to an agroforester to have permutation and combination of these species either with the agriculture crops, in form of agrisilviculture system, or in combination of other trees , in form of hortisilviculture systems. Further, there are lots of opportunities in finding suitable tree-crop combination as the sustainability and economic return of the system also varies with the aspect. A particular tree-crop combination which suited well for southern aspect will certainly not fit well with the climatic condition of the northern aspect at the same altitude. Similarly slope and altitude also have their role to play in finding out suitable tree-crop combinations. These all variations thus provide a huge opportunity for the agroforesters to work in Shivaliks.

Management of the system

Hills/ undulating lands provide a unique opportunity to work with tree-crop combinations. The farm lands in hills are in form of terraces whose width varies from few feet to several meters. Even the width of the terrace is not uniform throughout its length, as it tends to become narrower on either sides of the terraces. Further, the outer edge of the terraces, where the perennial components are planted, have many a times lesser soil depth, as the terraces are made by constructing the wall from loosely arranged stones. Such circumstances provide opportunity for standardizing the management practices (spacing, pruning, pollarding, thinning other management practices) so that the associated crops had minimum adverse effect of shading and nutrient competition.

Small and varying width of the terraces also provides a unique opportunity in finding suitable management practices. The width of the hedgerow or crown width of the tree need to be standardized for different location within the terrace so that the production of the crop do not vary drastically from centre of the terrace to either sides of the terrace and from outer edge to the inner side of the terrace.

Soil and water conservation

Shivaliks are the hot spot of erosion in the world. Soil erosion greater than 80 t $ha^{-1}yr^{-1}$ is being recorded from denuded hills of Shivaliks. To check this menace number of engineering measures like check dams, gabions and other structures are recommended. However, with passage of time the efficiency of these engineering structures decreases as the silt starts depositing. Thus they provide temporary protection from erosion. The need is to check erosion from the starting point itself. This is possible by covering the surface of the earth with vegetation. Trees and grasses in combination provide the best protection to the soils. As a thumb rule, silvopasture agroforestry systems are recommended for higher elevations where slopes are steep. At mid elevations, where slopes are medium, hortisilviculture, agrisilviculture, agrihorticulture systems are advised. At lower elevations or gentle slopes/ plains those agroforestry systems are advocated which have commercial motives behind it. For example Poplar and *Eucalyptus* based agrisilviculture systems where timber production is main objective or a system where trees are grown on the boundaries and main aim is agricultural crop production. Since, Shivaliks have undulating topography challenge to conserve soil and water is more typical. There is utmost need to research and recommend different tree-crop combinations which are capable of conserving resources of Shivaliks in different problematic areas. The altitude factor, slope factor, aspect factor and other physiographic factors needs to be taken care of while devising agroforestry systems for soil and water conservation measures.

Conclusion

Shivalik mountain ranges are unique land mass owing to their origin, vegetation and population pressure. Once covered with thick vegetation, they are now barren and have very high rate of soil erosion. The region is rainfed and hence agriculture is a risky proposition. In such situation land use systems which involves tree as one of the component can fit well to meet out the aspiration of the communities living in the region. Agroforestry not only provides fodder, fuel, timber, fiber but also allows crops to grow in combination. People of this region have been nurturing the trees on the farm since time immorial. However, the selection, and arrangement of trees are arbitrary. The need is to scientifically explore the variations in edaphic, topographic and climatic factors and accordingly develop the most suitable site specific tree-crop combination.

References

Agnihotri, Y. (2000). Climatic trends *vis-a-vis* crop planning in Shivalik foothills. *In*: FiftyYears of Research on Sustainable Resource Management in Shivaliks (eds. S.P.Mittal, R.K.Aggarwal and J.S.Samra), Central Soil and Water Conservation Research and Training Institute, Research Centre, Chandigarh, pp. 362-366.

Arya, S.L and J.S. Samra (2001). Revisiting watershed management institutions in Haryana Shivaliks, India. CSWCRTI, Research Centre-Chandigarh.

Kukal, S.S, H.S. Sur and S.S. Gill (1991). Factors responsible for soil erosion hazard in sub-montane Punjab, India. *Soil Use Manage.* 7: 38–44.

ICRAF (2006). World Agroforestry Centre, Southeast Asia web site. (http://www.worldagroforestrycentre.org/sea).

Tiwari, A.K, R.K. Aggarwal, Y. Agnihotri, Pawan Sharma, Pratap Singh, R.P. Yadav, V.K. Bhatt and Ram Prasad (2005). Vision–2025, CSWCRTI-Research Centre, Chandigarh, pp.23.

Sidhu, G.S., C.S. Walia, C.B. Sachdev, K.P.C. Rana, R.P. Dhankhar, S.P Singh and M. Velayuthem (2000). Soil resource of N-W Shivaliks for perspective land use planning. *In*: Fifty Years of Research on Sustainable Resource Management in Shivaliks (eds. S.P.Mittal, R.K.Aggarwal and J.S.Samra), Central Soil and Water Conservation Research and Training Institute, Research Centre, Chandigarh, pp. 23-34.

Yadav, R.P., R.K. Aggarwal, S.L. Arya, Pratap Singh, Ram Prasad, P. Bhattacharyya, A.K. Tiwari and M.K. Yadav (2004). Rainwater harvesting and recycling technology for sustainable production in small agricultural watershed - Johranpur. Bulletin No.T-50/C-11. CSWCRTI, Chandigarh.

Chapter-2

Agroforestry for Soil and Water Conservation in Jammu Shivaliks

SANDEEP SEHGAL AND MOHAMMED SALEEM
Division of Agroforestry, Sher-e-Kashmir University of Agricultural Sciences and Technology, Faculty of Agriculture, Chatha, P.O. Bhour Camp-180 009

ABSTRACT

Shivaliks are highly fragile ecosystem of Himalayas threatened by problems like soil erosion, loss of tree cover, decreased productivity of fuel, fodder and food, to name a few. These problems are further being complicated due to ever increasing anthropogenic pressure. Maintenance and enhancement of soil fertility is not only vital for food security and environmental sustainability in the Shivalik hills, but all over the globe. Agroforestry is frequently proposed as a solution to problems of land and water degradation as well as an answer to shortages of food, fodder and fuel in the Shivaliks, and the same holds true for Jammu Shivaliks. Substantial environmental benefits accrue from planting trees. Suitable tree crop combinations as well as their management practices need to be worked out. Growing of trees and grasses in agroforestry systems such as agrihorticulture, silvipastoral, agrisilvipastoral are specially suited for Shivalik hills keeping in view their role in soil and water conservation. Additional agroforestry component i.e, animals can further hasten the ameliorative effect of agroforestry. In all, combination of trees, crops and/or animals on the same unit of land holds great potential for contributing to sustainable land use system which can help in conserving soil and water besides providing innumerable socio-economic benefits.

Introduction

The Shivaliks are comprising of low hills form part of the outer Himalayas. They are distributed widely throughout the country. In the Western Himalayan region, the Shivalik foothill region is sandwiched between the Himalayas and the Indo-Gangetic alluvial plains and is spread over an area of 3 million hectares. The area is comparatively much lower in altitude and merges with the plains. It is spread in the states of Himachal Pradesh, Punjab, Jammu and Kashmir, Uttarakhand and Haryana and Union territory of Chandigarh. Locally, it is known as '*Kandi*'. In Jammu, Jammu & Kashmir, the Shivalik range lies to the south of *Pirpanjal*. The average height of Jammu Shivaliks is about 1000 m above mean sea level, which ranges from about 300 m to about 2000 m. The soil strata comprises of boulders and pebbles, gravel and sand with minor clays, mixed in varying proportions. Land has undulating topography, steep slopes with poor vegetative cover. The area has numerous streams which remain dry for most of the year, commonly called as '*Choes*' (torrents). Besides average rainfall of more than 1100 mm per annum, this region is still rainfed. Erratic distribution of rainfall often leads to frequent dry spells (Gupta *et al*, 1990).

Problems of Jammu Shivaliks

Like other Himalayan people, agriculture is the primary occupation of the people of Jammu Shivaliks. The people are dependent solely on forest ecosystems for food, fodder and fuel. However, during the last few decades all the natural resources in the Shivalik foothills are in a state of degraded condition. Primary reason responsible for this is the rising human as well as cattle population. Increase in number of animals per unit area leads to over grazing. People keep large herds of cows and goats and feed them by grazing on the surrounding hills. This heavy grazing does not allow the naturally growing grass to recover. It has been estimated that in the highly grazed Shivalik hills, as much as 6 cm of topsoil, representing nearly 2400 years of nature's work, can disappear in one year (Aggarwal *et al.*, 1987).

Similarly, the trees are cut/lopped specially, during the lean period when no other source of green fodder is available, making their survival/regeneration difficult. Rising population has also led to increased use of chemical fertilizers, unscientific farming and construction works. Increasing area is brought under agriculture including cultivation on slopes. All these factors lead to the loss of top soil. This surface soil is the richest part of the soil profile and is the most important factor determining the productivity and sustainability of all other natural resources. Soil erosion is high, particularly in regions with sparse vegetative cover. The dislodgement of soil particles from the surface using energy imparted to it by falling raindrops is a primary agent of erosion. The major result of the impulse imparted to the soil surface by raindrops is the breakdown of soil aggregates. The subsequent splashing of soil particles tends to puddle and close the soil surface, and thereby increase surface run-off. Surface run-off, combined with the beating action of raindrops causes rills to be formed on the soil surface. Rill erosion produces the greatest amount of soil loss worldwide (Singh and Phadke, 2006). Loss of this surface soil thus, poses economic and environmental concern (Gill, 1995; Dixon, 1997).

Role of Agroforestry

Since land is the most important natural resource of Shivaliks, therefore, an alternative, low cost approach is required, where both physical and socio-economic constraints can be overcome. Practices that minimize the rate of soil degradation, improve soil fertility, increase crop yields and increase farm income are keys to sustaining agricultural productivity. Out of a numerous options available, agroforestry has been proposed as a sustainable land use that can mitigate soil erosion and promote the economic welfare of small farmers. Keeping in view the sole dependence of the inhabitants on their surrounding forests, it is clear that any programme that checks or delays the flow of forest products to the people can not succeed. Since trees are an integral part of any agroforestry system, therefore, it paves way for the adoption of agroforestry in Shivaliks for the conservation of soil and water. Soil conservation is one of the primary benefits of agroforestry (Young, 1989). Agroforestry can be used as a tool for improving the land resources, economic viability and livelihoods especially of resource poor farmers. Agroforestry practices encompass an entire spectrum of land use in which woody components are deliberately used on the same land management unit as agricultural crops and or/animals, either in the same form of spatial arrangement or temporal sequence (Lundgren and Raintree, 1982). Another recent definition of agroforestry was proposed by Leakey. Leakey's (1996) definition is used most frequently: "A dynamic, ecologically based, natural resource management system that, through the integration of trees on farms and in the landscape, diversifies and sustains production for increased social, economic and ecological benefits". Trees are able to withstand adverse climatic conditions, thereby maintaining higher productivity as compared to agricultural crops. Trees are able to produce more fodder per unit area than agricultural crops. Besides fodder, trees also provide fuelwood when they are lopped so they help in meeting the energy requirements of local people. Although these are the common benefits of trees whenever they are integrated in an agroforestry system, but soil and water conservation benefits offered by the trees make them highly demandable for Shivaliks.

Trees have different impacts than annual crops on soil properties, because of their longer gestation period, larger biomass accumulation and continuous and more extensive root systems. Long gestation period of trees results in larger above and below ground biomass accumulation. It has been proved that trees can bring down the erosion rate appreciably. Leaf fall from trees form a thick layer of litter on the floor which decreases the impact of rainfall. Vegetation cover is important for soil protection because it reduces the kinetic energy of runoff and favours water infiltration in soil and plant roots hold the soil in position and protect it from being washed away. It breaks the impact of raindrops, decreasing their erosive capacity (Andreu *et al.*, 1998). This decomposed litter improves the soil structure, which decrease the surface run off volume. Furthermore, trees capture nutrients from the soil especially deeper layers and store them in their biomass thereby improving soil physical conditions. Trees like *Acacia senegal* and *Acacia albida* may halt or reverse soil degradation by soil fertility enhancement properties in semi-arid zone. *Proposis spp* have also demonstrated the capacity to improve soil fertility in semi-arid to dry savanna regions (Young, 1989). The perennial nature of tree root systems provides a dependable source of carbon substrate for microorganisms in the rhizosphere; microbial mucilage binds soil particles into stable aggregates, which results in improved soil structure (Sanchez, 2002). These processes take place continually in agroforestry systems instead of temporarily, as in agricultural systems (Lefroy *et al.*, 1999; Nair *et al.*, 1999; Sanchez *et al.*, 1997) and can be summarized as below:

- An increase in the soil organic matter content as a result of Carbon fixation in photosynthesis and its transfer via leaf fall and root turn over to soil.
- Leguminous trees fix atmospheric Nitrogen and result in an increase in soil N under the tree canopy.

- Rhizosphere effect of trees on soil resulting in enhanced N mineralization and increased microbial biomass.
- Microclimate modification by tree canopies that moderates soil and air temperatures and soil moisture regimes.
- Nutrient pumping reflecting uptake of nutrients from greater soil depths by tree roots and accumulation in a smaller volume of surface soil as a result of litter fall.

Efforts were made in the Jammu Shivaliks, to decelerate the soil erosion and improve the socio-economic conditions of the area. For this purpose, the Integrated Watershed Development Project (IWDP) was initiated. Under the Integrated Watershed Development Project (IWDP), Hills-II, efforts were made to rehabilitate the degraded Shivalik hills of Jammu, which covered two sub- watersheds in Shivaliks *viz.*, Ramnagar and Akhnoor. Revegetation of village common lands and grazing lands under hill and hillocks was done. Other common agroforestry interventions included the planting of fast growing trees and grasses, legumes and shrubs.

The benefits of trees cannot be overemphasized. Some common agroforestry trees suitable for Jammu Shivaliks is given in table 1.

Table 1 : Agroforestry trees suitable for Jammu Shivaliks

Tree species	Family	Common name	Uses
Adina cordifolia	Rubiaceae	Haldu	Fuel, Timber, Medicine
Ailanthus excelsa	Simaroubaceae	Mahrukh	Wood, Medicine
Aegle marmelos	Rutaceae	Bael	Fuel, Fruit, Medicine
Albizia lebbeck	Mimosaceae	Siris	Fuel, Timber, Nitrogen Fixing
Artocarpus lacucha	Moraceae	Dheu	Fodder, Fruit, Fuel
Bauhinia variegata	Caesalpiniaceae	Kachnar	Fodder, Fuel, Timber, Medicine, Nitrogen Fixing
Bombax ceiba L.	Bombacaceae	Semal	Fuel, Fibre, Timber, Medicine
Cassia fistula	Caesalpiniaceae	Kirala	Fodder, Fuel, Timber, Medicine, Nitrogen Fixing
Celtis australis	Ulmaceae	Khairk	Fodder, Fuel, Fruit, Timber, Medicine

Cordia dichotoma	Ehretiaceae	lasora	Fuel, Fruit, Timber
Emblica officinalis	Euphorbiaceae	Amla	Fruit, Medicine
Ficus spp	Moraceae	Fig	Fodder, Fuel, Timber, Medicine, Fibre
Grewia optiva	Tilliceae	Dhaman	Fodder, Fuel, Fruit, Fibre, Medicine
Melia azedarach	Meliaceae	Drek	Fodder, Fuel, Timber, Medicine
Moringa oleifera	Moringaceae	Sainjna	Fodder, Fuel, Medicine
Morus alba	Moraceae	Sahtoot	Fodder, Fuel, Fruit, Fibre, Medicine
Myrica esculenta	Myricaceae	Kaphal	Fuel, Fruit, Timber, Medicine
Pyrus pashia	Rosaceae	Kainth	Fuel, Fruit, Timber, Medicine
Quercus spp.	Fagaceae	Ban	Fodder, Fuel, Timber, Medicine
Syzygium cumini	Myrtaceae	Jamun	Fuel, Fruit, Timber, Medicine
Terminalia bellirica	Combretaceae	Bahera	Fodder, Fuel, Fruit, Timber, Medicine
Terminalia chebula	Combretaceae	Harad	Fodder, Fuel, Fruit, Timber, Medicine
Toona ciliata	Meliaceae	Tun	Fuel, Timber

In Jammu Shivaliks, the inclusion of trees and grasses resulted in decreased biotic pressure on the natural forests, reversed the trend of deforestation, and slowed down water run-off and increased the water regime of the catchments. It also resulted in an increase in the yield of different crops, milk, wool and meat, resulting in higher bio-physical gains (Sudan, 2009). A study in the western Himalayan valley region of India was carried out on the effect of different agroforestry systems *viz.*, block plantation and alley farming on runoff and soil loss with *Leucaena leucocephala* and *Eucalyptus* hybrid as the tree components and *Zea mays, Chrysopogon fulvus* grass or *Curcuma longa* as the crop component, on the runoff and soil losses. The reduction in erosion varied between 27 to 40 per cent in different treatments. It was primarily due to the barrier effect of tree or hedgerows and micro-terraces formed through sediment deposition along the contour barriers (Narain *et al.*, 1997). The study recommended the use of such vegetative measures that are productive while being protective for erosion control in areas with gentle slopes that are vulnerable to heavy erosion.

Some other studies (Prasad *et al.*, 2005) involving the suitability of different agroforestry systems for soil and water conservation in Shivaliks indicate that hortipastoral and silvopastoral systems are best suited agroforestry systems, considering the soil and climatic factors. Hortipastoral system involving amla+chrysopogon was most promising for productive and economic utilization and rehabilitation of degraded Shivalik foothills. A list of most adapted fruit species which may be successful on adverse agro-climatic conditions as well as degraded lands of Jammu Shivaliks is presented in Table 2.

Table 2 : Some fruit trees suitable for Jammu Shivaliks

S. No.	Common name	Scientific name
1	Mango	*Mangifera indica*
2	Guava	*Psidium guajava*
3	Citrus spp.	*Citrus lemon*
4	Pomegranate	*Punica granatum*
5	Ber	*Zyziphus mauritiana*
6	Amla	*Emblica officinalis*
7	Bael	*Aegle marmelos*

Conclusion

The Jammu Shivaliks, once dominated by forests, has rapidly deteriorated in the recent years because of the interaction of a number of interdependent factors affecting life and agriculture. Growing trees can play important role in conservation of soil and improving the overall land productivity of the Shivaliks. Tree based farming systems like agroforestry help to meet the subsistence requirements for food, fuelwood, and fodder. Agroforestry systems especially agrihorticulture and silvipastoral systems are crucial for the conservation of water and checking soil erosion.

References

Aggarwal, A., R.Chopra, and K Sharma. (1987). The State of India's Environment. The First Citizen's Report, Centre for Science and Environment, New Delhi, p 4.

Andreu V, J.L.Rubio and R.Cerni, (1998). Effects of Mediterranean shrub cover on water erosion. *Journal of Soil Water Conservation,* 53:112-120.

Dixon, J. (1997). Analysis and Management of Watersheds. *In* P. Dasgupta and K. Golan-Miiler (eds), The environment and emerging development issues, Volme 1. Clarendron Press, Oxford.

Gill, G. (1995). Major Natural Resource Management Concerns in South Asia. Food, Agriculture and the Environment Discussion Paper 8. International Food Policy Research Institute, Washington, DC.

Gupta, R.D., S.V. Bali, H. Singh, N.R. Khajuria and V.K. Koul. (1990). Ecology of Kandi belt of outer Himalayas of Jammu Division, Jammu and Kashmir state. In: S K Chadha (ed.) Himalayas Environment Problems.

Leakey R.R.B. (1996). Definition of agroforestry revisited. *Agroforestry Today,* 8(1): 5–7.

Lefroy, E. C.; R. J. Hobbs,.; M. H O'Connor and J. S. Pate (eds). (1999). Agriculture as a Mimic of Natural Ecosystems. Kluwer Academic Publishers, Dordrecht, The Netherlands.: 438.

Lundgren, B.O. and J.B.Raintree (1982). Sustained Agroforestry. *In* : Nestel, B. (Ed.). Agricultural Research for Development : Potential and challenges in Asia, ISNAR, The Hague, The Netharlands : 37-49.

Nair, P. K. R., R.J. Buresh., D.N. Mugendi, and C.R. Latt, (1999). Nutrient cycling in tropical agroforestry systems: Myths and science. pp.1-31. *In*: L. E. Buck, J.P. Lassoie, and E.C. Fernandes. (eds), Agroforestry in Sustainable Agricultural System,.CRC Press, Boca Raton, FL.

Narain Pratap, R. K. Singh, N. S. Sindhwal and P. Joshie. (1997). Agroforestry for soil and water conservation in the western Himalayan Valley Region of India 1. Runoff, soil and nutrient losses. *Agroforestry Systems,* 39(2): 175-189.

Prasad, R., R.K. Aggarwal, Y. Agnihotri, R.P. Yadav, J.S. Samra. (2005). Fruit based hortipastoral land use systems for the rehabilitation of degraded Shivalik foothills of north-west India. *Acta Horticulturae* 696: *VII International Symposium on Temperate Zone Fruits in the Tropics and Subtropics* - Part Two.

Sanchez, P.A., (2002). Soil fertility and hunger in Africa. *Science,* 295:2019–2020.

Sanchez, Pedro A., Roland J. Buresh Roger and R. B. Leakey (1997). Trees, Soils, and Food Security. Philosophical Transactions: *Biological Sciences,* 352 (1356): 949-960.

Singh Ram and V. S. Phadke. (2006). Assessing soil loss by water erosion in Jamni River Basin, Bundelkhand region, India, adopting universal soil loss equation using GIS. *Current Science,* 90 (10): 1431-1435.

Sudan, Falendra Kumar. (2009). Biophysical gains of participatory agroforestry: evidence from Integrated Watershed Development Project, Hills – II, India, Book of Abstracts, 2nd World Congress of Agroforestry. Agroforestry - The Future of Global Land Use, 23-28 August 2009, Nairobi, Kenya. pp.99.

Young A. (1989). Agroforestry for soil conservation. Wallingford, UK: CAB International.

Chapter-3

Livelihood Security through Agroforestry in Jammu Shivaliks

KAMAL KISHOR SOOD
Division of Agroforestry, Sher-e-Kashmir University of Agricultural Sciences and Technology, Main Campus, Chatha, Jammu (J & K)- 180 009

ABSTRACT

The land available to grow food and trees is decreasing owing to population pressure in all the regions of the world. In the Shivaliks, the land available for crop and tree cultivation is further limited by steep slopes and inaccessibility of different land areas. The livelihood options in the Jammu Shivaliks are limited due to poor development of this area compared to adjoining plains. In such a situation, agroforestry is one of the options to increase the livelihood security of the people of this region as it results in concurrent and sustainable use of the limited land for production of food and tree-based products and improves the livelihoods of the rural people. Agroforestry improvement based on scientific inputs and its popularisation has the potential to improve the livelihoods of people of this region. This paper dwells on the role of agroforestry in livelihood security and constraints in realising the potential of agroforestry in Jammu Shivaliks.

Introduction

The Shivaliks form a belt extending from south west Kashmir through northern India to Nepal. These are known as sub-Himalayas.

These extend through Jammu (J & K), Kangra valley (Himachal Pradesh), Sirmour (Himachal Pradesh) to Dehradun (Uttarakhand) through the state of Haryana and, Garhwal and Kumaon in Uttrakhand (Anonymous, undated). A part of the Shivaliks lies in Jammu and Kashmir province of India. The state of Jammu and Kashmir has a forest cover of 22886 km^2 which forms only 10.2% of geoegraphical area of the state (FSI, 2009) as against two-third of the geographical area envisaged in the Indian National Forest Policy of 1988. The Shivaliks of Jammu constitute semi-arid and sub-humid climate (Sharma *et al.* 2004). A considerable part of Shivaliks have rugged and undulating topography and, inaccessibility and poor development (Kaul *et al.*, 1991). A major part of this region is *Kandi* with stony and sandy soil strata (Gupta *et al.*, 1990). The people of this region have narrow options for their livelihood security. Rainfed subsistence agriculture is the main stay of the people and they are dependent on state forests for fuelwood, domestic timber, fodder and other tree based non-timber products (Kaul *et al.*, 1991). Increase in human population coupled with their dependence on forests has resulted in degradation of the forests in this region. This has, consequently, resulted in the livelihood insecurity for want of availability of sufficient tree products and concomitant ecological hazards. Agroforestry, tree cultivation on farm lands in various forms, is one of the ways to provide tree based products to farmers on their door steps, reduce the pressure on the existing forests and provide associated tangible and intangible benefits at local, regional and global levels.

Role of Agroforestry for Livelihood Security

A livelihood comprises the capabilities, assets (including material and social resources) and activities required for living (Carney, 1998; Ellis, 1998). A livelihood is more than just income (cash or kind). It includes: income (cash or kind), social institutions (like family, kin, village), gender relations and property rights required to support and sustain a given standard of living (Ellis, 1998; Kragten *et al.*, 2001). A livelihood also includes access to and benefits derived from social and public services provided by the state such as education, health, roads and water supply (Ellis, 1998). A livelihood is sustainable and secure when it can cope with and recover from stresses and shocks and maintain or enhance its capabilities and assets both in the present and future, while not undermining the natural resource base. Livelihood strategies are the ways in which households try to sustain or improve their livelihoods. According to theory of livelihood security, farming households pursue the following objectives (Arnold, 1997; Carney, 1998; Ellis, 1998; Kragten *et al.*, 2001; Scherr, 1995):

- Secure provision of food and essential subsistence goods
- Cash for purchase of outside goods and services
- Savings (resource accumulated to meet future planned needs and emergencies).
- Reduce vulnerability (reducing critical risk factors)

Trees in agroforestry systems have many important roles in livelihood security of the people in general and more specifically in the Shivalik belt of Jammu, which is pre-dominantly rainfed with undulating and rugged topography and degraded soils. The farmers of this region are poor owing to narrow livelihood options. Like the remaining Himalayas, the Shivaliks are characterised by fragility, poverty, inaccessibility, marginality (resource scarcities and uncertainties) and harsh climate. In such situations, agroforestry has important role in meeting the livelihood objectives of the farmer so that they can achieve livelihood security (Sood, 2009). The roles of the on-farm tree cultivation (agroforestry systems) in the livelihoods of the farmers in the Shivaliks are delineated in Table 1.

Table 1 : Agroforestry for livelihood security in Jammu Shivaliks

Livelihood Objectives	Role of agroforestry in livelihood security in Jammu Shivaliks
Secure provision of food and essential subsistence goods	• Agroforestry systems not only produce food through crops, but also the trees directly provide food in the form of fruits, vegetables and also act as nutrient supplement to the farmers. The fruits like *Ficus roxburghii, Sygygium cumini, Juglans regia, Pyrus pashia, Morus spp., Prunus spp.* are eaten by people which provide additional minerals and calories. • Trees improve the soil which increases food productivity resulting in food security. • Tree growing in agroforestry systems provide subsistence goods and inputs (implements, ploughs, tool handles etc.) for farming. • Many trees in Shivaliks provide green fodder at the times of fodder scarcity e.g. *Grewia optiva* and *Bauhina variegata*.

Cash for purchase of outside goods and services	• Tree products growing in on-farm agroforestry systems are sold by the farmers which provide them the cash income. This in turn help the poor farmers to buy outside goods and services required for their livelihood security e.g. sale of tree products from many species like *Syzygium cumini, Toona ciliata, Albizzia lebbeck, Ficus roxburghii, Dendrocalamus spp., Terminalia spp* etc. provides considerable amounts of household income.
Savings	• Trees grown on the farms act as a saving which can be used or sold to meet household planned needs or exigencies or other emergencies.
Reduce vulnerability	• Agroforestry helps to restore nutrient in the upper layer of the soil and enhances the site productivity. • Trees growing in agroforestry systems, provide income in times of poor crop production, unemployment and other kinds of hardships, contingencies or to meet exceptional needs. • Trees help out to utilise excess household labour in activities like tree planting, management, harvesting of tree products etc. • Agroforestry systems provide biomass reserve to cook food. • On-farm agroforestry trees, reduces the burden on the household labour to bring fuelwood loads from the state forests. • People in Shivaliks use many trees products to cure ailments e.g. *Terminalia belirica, Terminalia chebula, Embilca officinalis, Syzygium cumini.* This reduces the health risks to the households. • Helps in conservation of biodiversity

Based on: Sood, 2003

Major Types of Agroforestry Practices in Jammu Shivaliks

1. Tree cultivation in homegardens
2. Trees on cropland (as intercrop)
3. Tree on bunds of cropland
4. Trees on boundaries of agricultural holdings
5. Trees cultivation on pasture/grassland

The important agroforestry trees for Jammu Shivaliks are listed in Table 2.

Table 2 : Trees Suitable for Agroforestry Systems in Jammu Shivaliks

Name of species	
Acacia catechu	*Leucaena lecocephala*
Acacia modesta	*Morus spp*
Acacia nilotica	*Populus deltoids*
Albizzia lebbeck	*Prunus cerasoides*
Aegle marmelos	*Pyrus pashia*
Ailanthus excelsa	*Quercus luecotrichophora*
Bauhinia variegata	*Quercus dilatata*
Bombax ceiba	*Salix spp*
Butea monosperma	*Sapindus spp.*
Cassia fistula	*Rhododendron spp.*
Celtis australis	*Syzygium cumini*
Cinnamomum tamala	*Terminalia bellirica*
Dalbergia sissoo	*Terminalia chebula*
Dendrocalamus spp	*Terminalia tomentosa*
Emblica officinalis	*Terminalia arjuna*
Eucalyptus spp	*Robinia pseudacacia*
Ficus palmata	*Toona ciliata*
Ficus roxburghii	*Pinus roxburghii*
Grewia optiva	*Pinus wallichina*
Grevillea robusta	*Ziziphus mauratiana*

Constraints in Realising the Potential of Agroforestry in Shivaliks

The following are important constraints in realising the potential of agroforestry systems in livelihood security of the farmers in the Jammu Shivaliks

- Lack of availability of quality planting stock of agroforestry tree species
- Lack of awareness about suitable tree-crop combinations
- Lack of know-how of scientific agroforestry
- Long gestation period of trees to yield suitable tree product and poor efforts to produce prototypes yielding in shorter duration
- Conflicting messages from different line departments about the trees

- Legal and procedural impediments in harvesting, transport and sale of tree products
- Poor knowledge on value addition and storage of tree products
- Non-existence of formal markets for sale of tree products

Conclusion

There is need to popularise agroforestry systems in Jammu region on scientific lines and provide better quality seedlings to the farmers for getting overall higher yield and benefits from the agroforestry systems in Jammu Shivaliks so that roles of trees in farmers' livelihood security can be improved. The lifting of legal restrictions on harvesting, transport and sale of tree products from agroforestry systems is needed for better adoption of agroforestry systems in this region.

References

Anonymous. (undated). Shivalik hills. http://www.indianetzone.com/4/shivalik_hills.htm (accessed on 23.01.10).

Arnold, J.E.M. (1997). Social dimensions of forestry's contribution to sustainable development. *Proceedings of the XI World Forestry Congress, Antalya, Turkey, October 12-22*. Volume 5. pp. 3-23.

Ellis, F. (1998). Household strategies and rural livelihood diversification. *The Journal of Development Studies* 35: 1-38.

Carney, D. (1998). Implementing the sustainable rural livelihoods approach. In: Carney, D. (Ed) *Sustainable Rural Livelihoods. What Contribution Can We Make*? DFID. pp. 3-23.

FSI (2009). Forest Survey of India, *State of Forest Report 2009*.(http://www.fsi.nic.in/sfr_2009/chapter2.pdf)

Kaul, V.K., R.D Gupta, H. Singh, and M. Saleem, (1991). Agroforestry as an alternative land use management system in *Kandi* belt of Shivalik Himalayas of Jammu region, Jammu and Kashmir state. *In*: Gupta, R.D. and Benerjee, S.K. (Eds.). *Problems and Management of Soil and Forest Resources of North Western Himalayas*. Mahajan Book Centre, Last morh, Gandhi Nagar, Jammu. pp. 147-156.

Kragten, M., T.P.Tomich, S.Vosti, and J. Gockowski, (2001). Evaluating land use systems from a socio-economic perspective. *ASB Lecture Note 8*. International Centre For Research in Agroforestry, Bogor, Indonesia. pp. 31.

Gupta, R.D., S.V. Bali, H. Singh, N.R. Khajuria, and V.K.Koul, (1990). Ecology of kandi belt of outer Himalayas of Jammu division, Jammu and Kashmir. *In*:

S.K. Chadha (Ed.). *Himalayas Environmental Problems*. Ashish Publishing House, Punjabi Bagh, New Delhi.

Sharma, V., S.Arora, V.K. Jalali, and D. Kher, (2004). Studies on constraints in adoption of improved soil and rainwater conservation practices in Kandi region of Jammu. *Journal of Research, SKUAST-J*. 3(2): 214-220.

Scherr, S.J. (1995). Economic analysis of agroforestry systems: The farmers' perspective. *In:* Current, D., Lutz, E., and Scherr, S (Eds.) *Costs, Benefits, and Farmers Adoption of Agroforestry*. World Bank Paper Number 14. pp. 29-44.

Sood, K.K. (2003). *Factors Affecting Tree Growing in Traditional Agroforestry Systems in Western Himalaya, India*. PhD Thesis, University of Aberdeen. Aberdeen, United Kingdom.

Sood, K.K. (2009). Agroforestry for sustainable development of watershed and livelihood security. In: Gupta, N.K., Gupta, L.M., Bali, A.S. (Eds.). *14 days Training of Trainers on Scaling Up of Water Productivity in Agriculture for Livelihood through Teaching cum Demonstration (18-31 August, 2009)*. Water Management Research Centre and Directorate of Research, SKUAST-Jammu. pp. 101-105.

Chapter-4

Soil and Nutrient Loss under Different Land-Uses in Upper Shivaliks of Himachal Pradesh

I P SHARMA AND SUDHIR VERMA
Department of Soil Science and Water Management, College of Forestry, Dr. Y S Parmar University of Horticulture and Forestry, Nauni-173 230, Solan (HP)

ABSTRACT

Shivaliks, the most important ecosystem of the foothills of the Himalayas, have been identified as one of the eight most degraded rainfed ecosystems of the country. They are characterized by undulating barren hill slopes, undulating agriculture fields and erratic rainfall pattern. Majority of the lands under cultivation and orchards in the hilly terrain of the state are thus subjected to runoff and soil loss of varying degrees. The soils of Shivalik hills in Solan district of Himachal Pradesh were found to be coarse in texture with abundance of gravels (upto 73.20%). According to the erodibility indices i.e erosion ratio (ER) and erosion index (EI), all the soils were found to be erodible; however, grasslands and forestlands recorded lowest values for these indices. Forest soils had higher water stable aggregates (WSA), mean weight diameter (MWD) and lower dispersion ratios than the cultivated and orchard soils. The dispersion ratio was comparatively higher under cultivated lands (17.37-22.54) followed by orchards (14.36-14.99) and lowest under forest soils (12.59-13.70). Areas having barren lands recorded very high soil loss (86.05 t ha^{-1} yr^{-1}) followed by intensively cultivated areas (58.87 t ha^{-1} yr^{-1}) and both of these have fallen in very high category of priority. A medium category of soil loss has been observed under degraded forests, agro-horticultural plantations and scrubs (14.77 - 18.27 t ha^{-1} yr^{-1}). While low to very low soil loss was recorded under moderately dense, mixed and dense forests (3.46 – 9.59 t ha^{-1} yr^{-1}). Weighted soil loss of 10.60, 17.12, 19.20 and 256.73 t ha^{-1} yr^{-1} was observed under 5-10, 10-15, 15-35 and >35 per cent slope classes. However, dense forests in all the slope classes registered annual soil loss within the tolerance limit (5 t ha^{-1} yr^{-1}). Hence, maximum runoff, soil loss and available NPK losses were recorded from the barren lands and lands under cultivation followed by orchards and minimum under forests. Risk of soil erosion is higher on steeper slopes. Mixed forests in slope class of 15-35% have also reflected very high soil loss. However, dense forests in all the slope classes registered annual soil loss within the tolerance limit (5 t ha^{-1} yr^{-1}). There is an urgent need to adopt suitable soil and water conservation measures to check runoff losses causing loss of soil and nutrients.

Introduction

Soils are the most valuable and vital natural resource for the socio-economic development of a country. In India, out of the total geographical area of 328 m ha, about 147 m ha is reported to be suffering from soil degradation processes of various types (NBSSLUP, 2004). Soil erosion is a major cause of soil quality deterioration. The increased soil erosion takes away productive top soil, forcing decline in crop yields and tremendous offsite or downstream impacts. It has been estimated that more than 5.3 billion tones of top soil is being eroded every year, at an average rate of 16.3 t ha^{-1} yr^{-1}, out of which 1.6 billion tones representing about 29% of total eroded mass is permanently getting lost to the sea and about 10% gets deposited in the reservoirs, reducing their capacity by 1-2 % every year (Dhruvanarayana and Ram Babu, 1983; Singh *et al.*, 1992; Narain *et al.*, 1993).

Shivaliks, the most important ecosystem of the foothills of the Himalayas, have been identified as one of the eight most degraded rainfed ecosystems of the country. The Shivalik hill region lies between $30^0$10′ to $33^0$37′ N latitudes and $73^0$37′ to $77^0$39′ E longitudes and the foothills of the region covering an area of 3.08 m ha is in the states of Himachal Pradesh, Punjab, Haryana, Uttarakhand and Jammu and Kashmir (Mittal *et al.*, 2000; Yadav *et al.*, 2005) and out of this 1.14 m ha is in Himachal Pradesh. It constitute sub-montane, low hills and foothills sub-tropical area (lower Shivaliks) between 350-650 m amsl, and mid-hills sub-humid areas (upper Shivaliks) between 650-1820 m amsl in J&K, Punjab, HP, Haryana and Uttarakhand. The geological formation of Shivalik belt is very weak and vegetation is scanty as a result of which the area is subjected to high erosion. The soil of these areas is mostly sandy loam in texture with scattered loamy patches. The moisture retention capacity is very poor. Severe erosion (>80 t ha^{-1} yr^{-1}) has been reported from Shivalik hills (Singh *et al.*, 1992).

The denudation of fragile area of Shivalik started about 150 years ago and continued unabated in spite of legislative and control measures (Mishra and Sarin, 1987). As a result, area once protected with thick vegetation got converted into bush lands of little economic importance (Bhumbla, 1976). Due to excessive biotic pressure in these areas for fuel wood, fodder and timber, grazing requirements for livestock and developmental activities, forest resources are depleting at a very fast rate resulting in increased rate of siltation, surface runoff, sheet erosion and gully erosion. Excessive exploitation is much beyond the carrying capacity, and is no more sustainable. This has resulted in the depletion, degradation and endangering natural vegetation of the forests which has become a cause of serious concern.

In Himachal Pradesh also, about 75.4 per cent of the area is suffering from various forms of soil degradation (NBSSLUP, 2004). The soil erosion is a major problem of Himachal Pradesh especially in the sub-montane mid-hill zone (Upper Shivaliks). Majority of the lands under cultivation and orchards in the hilly terrain of the state are thus subjected to runoff and soil loss of varying degrees. This soil loss is rendering the land unfit for cultivation and simultaneously the low lying areas, water bodies, dams etc are also being affected due to deposition and/or siltation of this eroded soil. Information on the existing or potential erosion prone areas and landuse practices helps in identifying the location and extent of the sediment sources in an area and in taking up soil conservation measures on priority basis. An attempt has been made in this article to compile the work done on soil and nutrient loss under different land-uses in the upper Shivaliks of Solan district in Himachal Pradesh.

Study Area

The different studies on erodibility characterization and soil and nutrient loss, compiled in this article, were undertaken in Solan district of Himachal Pradesh located between $30^0 55'$ to $31^0 06'$ N latitude and $77^0 06'$ to $77^0 13'$ E longitude. The climate of the study area is sub-temperate sub-humid with mean annual temperature of 20^0C and annual rainfall of 1100 mm. Nearly 80 per cent rainfall is received during the rainy season. About 80% of the area is rainfed. Maize and wheat are the main crops. However, vegetables, sub-tropical fruits, pulses and oilseed crops are also grown.

One of the studies on soil loss was also conducted in Kawal khad watershed (part of Giri river catchment) which is located in Solan and Sirmour districts of Himachal Pradesh between 30°45′00″ to 30°54′45″ N latitude and 77°03′15″ to 77°13′35″ E longitude with an elevation ranging from 900 to 2100 m above msl. It covers an area of 165.2 sq km and is situated on the right bank of Giri river. The peculiar characteristics of this watershed are hilly terrain, steep slopes, intensive cultivation, variable soils, different types of landuse/landcover, excessive monsoon rains, heavy biotic interferences etc. It has sub-humid sub-temperate climate with mean annual rainfall of 1300 mm and mean annual temperature of 20°C. The soil moisture and temperature regimes are characterized by udic and thermic types, respectively. The soils of the watershed have developed from ferromagnesian shales, and dolomitic limestone. The major landuse categories found in the study area comprise of chir pine (*Pinus roxburghii* Sargent) forests, agrihorticultural plantation, intensive cultivation, scrubs and barren lands. Larger area of the watershed is under forests. Dense (>40% crop canopy), moderately dense (10-40% canopy), degraded (<10%) and mixed forests (variable canopy) covered 16.19, 20.95, 6.81 and 8.27 per cent watershed area, respectively. Agriculture (Agrihorticultural plantation and intensive cultivation), scrubs and barren lands covered 31.95, 12.73 and 2.11 per cent area of the watershed, respectively. The settlements and river covered a very small area (1.16%).

Material and Methods

Soil erodibility characterization considerably helps in developing sound management practices for reduction of the land degradation. Field investigation was undertaken in the mid hill region of Solan district to study various soil erodibility indices under different land uses (Kumar *et al.*, 2002). Study area comprised of eight locations and three land-uses *viz.* cultivation, orchards and forests. In the cultivated landuse, the land was fallow from June to Oct. 1996. In Nov., 1996 Pea crop was grown at 60 cm x 10 cm spacing following recommended practices. Under orchard land use, apricot orchard spaced at 8 m x 8 m was selected. The orchard was equipped with drip irrigation and basins were maintained weed free. Under forest land use, pure chir pine forest was selected. There were 10 yr old trees having more than 40 % canopy. Dispersion and erosion ratios were determined as per procedures outlined by Middleton (1930). Sharma *et al.* (2000) assessed the runoff, soil and nutrient losses from runoff plots under these three landuses. The plot was 10 m x 10 m in size with a uniform slope of 2 per cent. Each plot was bounded on all sides and equipped with 'V'-notch. The pre-calibrated runoff collecting drums were used to measure runoff after each rainfall event.

Sharma *et al.* (2002) and Sharma *et al.* (2003) studied the effect of watershed characteristics (such as landuse/land-cover, slope, etc.) on soil erosion in Kawal khad watershed using remote sensing and GIS techniques. Quantitative soil loss was computed for different landuse/landcover classes, soil types and slope classes. Landuse/landcover, physiography-cum-soilscape information obtained from false colour composite (FCC) of IRS-1C LISS-III imagery and terrain slope information obtained from topographic map and rainfall data were used to provide inputs to universal soil loss equation (USLE) of Wischmeier and Smith (1978).

Results

The soils of Shivalik hills were found to be coarse in texture with abundance of gravels (upto 73.20%). Soils were characterized by high bulk density and low porosity, due to their mechanical composition. These soils were found to be poor in moisture retention, highly permeable and had low available water capacity. Soil reaction was found to vary from acidic to slightly alkaline. Organic carbon was highest in forest soils followed by orchards and cultivated soils. Fertility status of the soils indicated that these were low in available N, high in available P and low to medium in available K status. The moisture retention characteristic at various suctions was found to be the highest in forest soils and the lowest in cultivated soils while the orchard soils were intermediate in moisture retention (Kumar, 1997; Kumar *et al.*, 2002; Kumar, 2005).

Erodibility indices

The dispersion ratio and erosion ratio have been used by various workers to assess the erosion behaviour of soils. Kumar *et al.* (2002) reported that dispersion ratio was comparatively higher under cultivated lands (18.50-21.82) followed by orchards (13.25-14.46) and lowest under forest soils (11.64-12.93) (Table 1). The erosion ratio followed similar trend. It was highest under cultivated lands (17.78-21.14) followed by orchards (13.42-15.50) and lowest under forest soils (13.24-14.38). The dispersion ratio was found to be significantly and positively correlated with erosion ratio (r=0.95). However, both dispersion ratio and erosion ratio showed significant and negative correlation with organic carbon. The soils under forest land use were more stable as compared to the soils under cultivation and it could be ascribed to higher OC resulting in higher water stable aggregates and higher MWD. The erosion behaviour was assessed by erodibility indices i.e dispersion ratio and erosion ratio which were well correlated (0.9564*) with each other and were found to be negatively influenced by the cementing agents like organic carbon (-0.6816* and - 0.6863*) and sesquioxides (-0.4214* and - 0.3767*), respectively and non-significantly with clay. On the other hand, stability indices i.e water stable aggregates (>0.25 mm) and mean weight diameter were positively and significantly correlated with organic carbon (0.8819* and 0.7333*) and sesquioxides (0.4533* and 0.4325*). From these findings it was revealed that the effect of organic carbon was more pronounced than other soil binding agents. According to above erodibility indices, all the studied soils are erodible as their dispersion ratio and erosion ratio are greater than 15 and 10, respectively. Among the land uses the forest soils were comparatively less erodible than orchard following by soil under cultivation. While studying the erodibility characteristics of soils under different land uses *viz.* agriculture, forest, grasslands, scrub and barren lands, Kumar (2005) also found that all the soils were erodible (based on erosion ratio); however, grasslands and forestlands recorded lowest values for the indices (Table 1).

Various soil characteristics alongwith topography, vegetation and biotic interference provide fairly good idea about the erodibility. It has been observed that soils under similar conditions of topography, climate and vegetative cover, differ naturally in the extent of their erosion. Such differences are due to physical make up of soils. The physical properties of the soils are not easily modified. In areas having similar climate but different land use/land cover, the variations in erodibility of the soils might be due to physical properties of the soils like texture, water stable aggregates, etc. in addition to topography and other inherent properties.

Table 1 : Erodibility characteristics of soils under different land uses.

Location (Land use)	Dispersion ratio	Erosion ratio	WSA (>0.25mm dia)	MWD (mm)
Nauni (C)	18.92	17.78	72.11	2.70
Nauni (O)	14.46	13.42	89.17	3.86
Lavighat (C)	20.42	21.14	68.36	1.73
Mukari (C)	21.82	20.80	66.24	1.25
Rundon Ghonron (C)	18.50	19.92	79.32	2.80
Majhat (F)	12.93	14.38	93.27	5.49
Kandaghat (O)	13.25	15.50	92.89	4.26
Ranno (F)	11.64	13.24	95.32	5.22

C = cultivation, F = forest, O = orchard
Kumar *et al.*, 2002

Soil and nutrient loss

The soil erosion takes away the productive top soil and thus impairs the soil health. It is, therefore, indispensable to assess the soil loss under different land-uses, prioritize the areas and devise suitable conservation measures to maintain and enhance the productivity.

Sharma *et al.* (2000) while assessing runoff and soil losses under three landuses *viz.* cultivated, orchards and forests revealed that forest soils were more stable than the soils under orchards and cultivation. The maximum runoff (13.5 % of the total rainfall) and soil loss (32.11 t ha^{-1}) were recorded from the lands under cultivation followed by orchards and minimum under forests (Figure 1). Of the total rainfall (1137.6 mm) received during the study period, 71.2 % (809.6 mm) was recorded during the rainy season i.e. June to Sept., 153.6 (13.5%), 100.7 (8.85%) and 65.4 mm (5.74%) was recorded as runoff under three landuses *viz.* cultivation, orchards and forests, respectively. Runoff losses were comparatively low in the month of June than in September. This could be due to the fact that in the month of June soils were invariably dry and a good amount of rainfall was stored in profile or lost in evaporation. Higher runoff losses during July to September could also be attributed to higher intensity of rainfall. Total soil loss was 32.11 t ha^{-1} under cultivated lands, 20.00 t ha^{-1} under orchards and 11.68 t ha^{-1} under forest lands. Soil loss during rainy season was

30.37, 18.93 and 10.93 t ha^{-1} under cultivation, orchards and forests, respectively accounting for 94.60, 94.55 and 93.50 per cent of the total annual soil loss under these landuse categories (Figure 1).

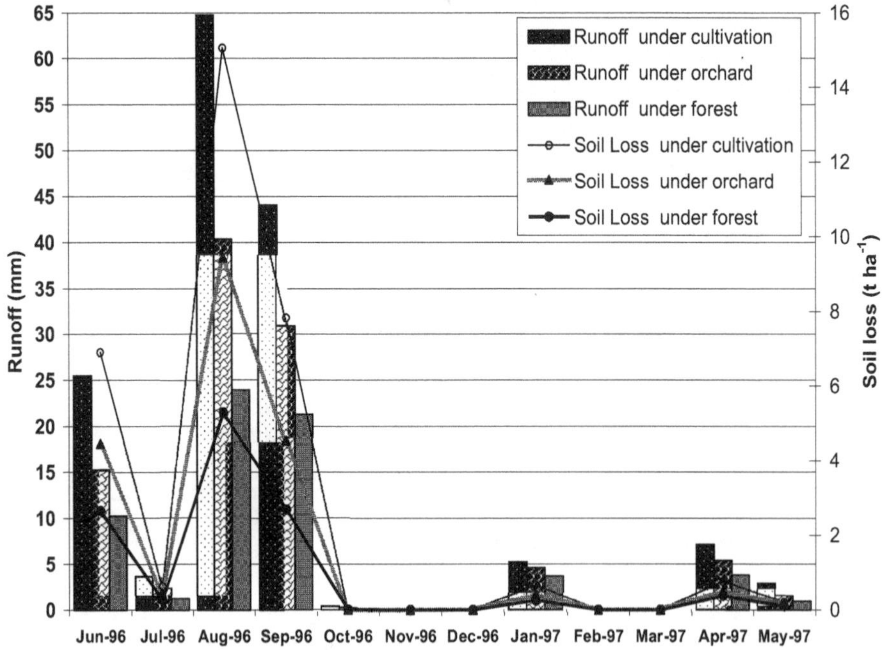

Fig. 1 : Runoff and soil losses under different landuses
Source : Sharma *et al.*, 2000

The soil properties, land use/land cover are important factors affecting sediment production (Wood, 1987). Both runoff and soil loss were comparatively higher under cultivated lands followed by orchards, and least under forest lands (Sharma *et al.*, 2000). This could be ascribed to better vegetative covers under orchards and forests resulting in interception of falling rain drops and preventing the direct beating action and also due to beneficial effect of vegetation on soil properties. Of the three landuse classes, forest soils had highest infiltration rate (5.91 cm hr^{-1}) as compared to cultivated (1.08 cm hr^{-1}) and orchards (2.90 cm hr^{-1}). The soils under forests also had well aggregated top soil layer resulting in higher infiltration, less detachment and migration of soil particles through runoff. Dhruvanarayana *et al.* (1986) observed that runoff was 50% of the rainfall in the denuded Shivalik hills and 36% in cultivated areas.

Sharma *et al.* (2002 and 2003) studied the effect of watershed characteristics on soil erosion in southern Himachal Pradesh using remote sensing and GIS techniques. They computed quantitative soil loss for different landuse/landcover classes, soil types and slope classes using universal soil loss equation (USLE) of Wischmeier and Smith (1978). The data on soil loss under different land uses (Table 2) revealed that watershed areas having barren lands recorded very high soil loss (86.05 t ha^{-1} yr^{-1}) followed by intensively cultivated areas (58.87 t ha^{-1} yr^{-1}) and both of these have fallen in very high category of priority. A medium category of soil loss has been observed under degraded forests, agrihorticultural plantations and scrubs (14.77 – 18.27 t ha^{-1} yr^{-1}). While low to very low soil loss was recorded under moderately dense, mixed and dense forests (3.46 - 9.59 t ha^{-1} yr^{-1}). Slope has a direct influence on soil and land use. It plays an important role with reference to the susceptibility of an area to soil erosion (Morgan, 1979). This could be well explained with the help of 'LS' factor which increases with the increase in slope and slope length. Moderately dense forests in slope class of 15-35 per cent also reflected very high soil loss. Weighted soil loss of 10.60, 17.12, 19.20 and 256.73 t ha^{-1} yr^{-1} was observed under 5-10, 10-15, 15-35 and >35 per cent slope classes. Thus, maximum soil loss was reported from slope class >35 per cent. However, dense forests in all the slope classes registered annual soil loss within the tolerance limit (5 t ha^{-1} yr^{-1}). Vegetation protects the soil against impact of falling rain drops, increases the roughness of the soil surface, reduces the speed of surface runoff, binds the soil mechanically and improves the physical, chemical and biological properties of the soil (Zatcher, 1982). Higher soil loss under cultivated land could be due to poor land cover. In addition the soils of the area are lighter in texture and weak in structure (Table 2).

Grewal (1993) also reported conservation of soil and water through alternative land use systems. He observed that silvipastoral and agrisilviculture systems provided best mechanism for conservation in Shivaliks. The soil loss and runoff under *Eucalyptus territicornis* plus *Eulaliopsis binata* reduced to 0.07 t ha^{-1} and 0.05 per cent in comparison to 5.65 t ha^{-1} and 23.0 per cent under cultivated fallow and 2.69 t ha^{-1} and 20.50 per cent under *Sesamum indicum-Brassica campertris* systems, respectively. Thus tree based systems were found to be very effective in Shivalik region where erosion is a major cause of soil degradation and nutrient depletion.

Table 2 : Effect of land use and slope on soil loss

Landuse class	Soil loss (thousand tones) Slope class (%)				Total	Area (ha)	Area weighted soil loss (t ha^{-1} yr^{-1})
	5-10	10-15	15-35	>35			
Dense forest	0.01	5.43	2.31	1.49	9.24	2674.15	3.46
Mixed forests	0.05	1.81	3.85	6.38	12.09	1366.29	8.85
Moderately dense forest	0.02	2.92	21.25	9.01	33.20	3461.14	9.59
Scrubs	0.00	7.87	3.45	19.74	31.06	2103.31	14.77
Degraded forests	0.00	5.44	7.66	6.88	19.98	1124.86	17.76
Agro-horticultural plantations	0.07	8.86	13.00	26.66	48.59	2659.33	18.27
Intensive cultivation	1.73	59.77	38.12	54.47	154.09	2617.33	58.87
Barren lands	0.00	6.14	11.16	12.76	30.06	349.34	86.05
Soil loss (th ton)	1.88	98.24	100.80	137.39	338.31	-	-
Area (ha)	177.40	5737.00	5250.40	535.15	-	-	-
Area weighted soil loss (t ha^{-1} yr^{-1})	10.60	17.12	19.20	256.73	-	-	-

Sharma *et al.*, 2002

The loss of top soil through erosion also results in nutrient loss, thus forcing a decline in crop productivity. Sharma *et al.* (2000) estimated nutrient loss in the runoff water during June 1996 - May 1997 and observed that the loss of available N, P and K was 58.47, 16.46 and 10.62 kg ha^{-1} from cultivated lands, 28.22, 7.20 and 4.58 kg ha^{-1} from orchard soils and 10.74, 3.58 and 2.13 kg ha^{-1} from forest lands (Figure 2a,b). Under these three landuse systems *viz.* cultivation, orchards and forests was more than 95% of the nutrient loss was during rainy season. The loss of nutrients was highest from cultivated lands followed by orchards and lowest under forested areas. The trend in nutrient loss with time was similar to that observed for runoff and soil loss. Of the three nutrients, N loss was found to be the highest in all the landuse classes and this could be ascribed to more solubility and less absorption of different forms of N by soil colloids resulting in more losses through runoff (Kale *et al.*, 1992; Mathan and Kannan, 1993).

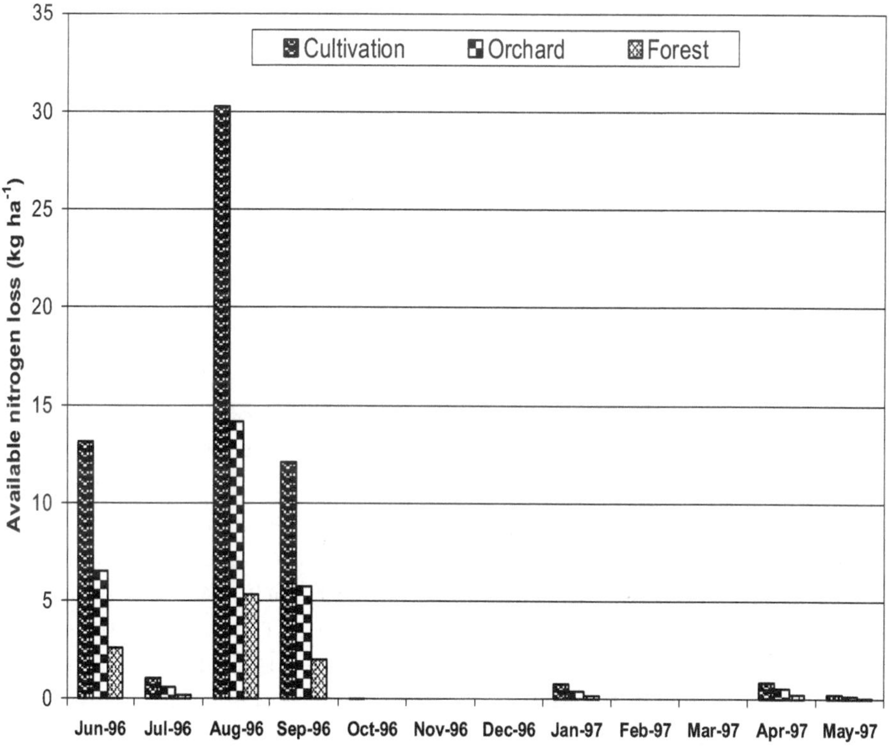

Fig. 2a: Available N, P & K losses (kg ha^{-1}) in sediment under different landuses
Source : Sharma *et al.*, 2000

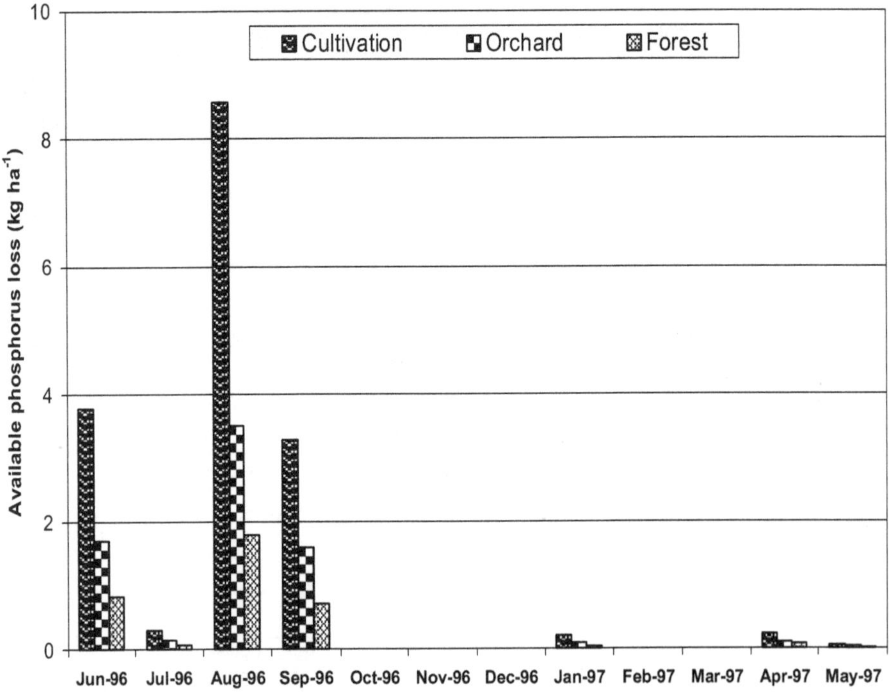

Fig. 2b : Available N, P & K losses (kg ha^{-1}) in sediment under different landuses
Source : Sharma *et al.*, 2000

Conclusion

Shivaliks are characterized by undulating barren hill slopes, undulating agriculture fields and erratic rainfall pattern. Majority of the lands under cultivation and orchards in the hilly terrain of the state are thus subjected to runoff and soil loss of varying degrees. Maximum runoff, soil loss and available NPK losses were recorded from the barren lands and lands under cultivation followed by orchards and minimum under forests. The soil loss varied from 3.46 – 9.59 t ha^{-1} yr^{-1} under moderately dense, mixed and dense forests; 14.77 – 18.27 t ha^{-1} yr^{-1} under degraded forests, agrihorticultural plantations and scrubs; 58.87 t ha^{-1} yr^{-1} under intensively cultivated areas; and 86.05 t ha^{-1} yr^{-1} under barren lands. Risk of soil erosion is higher on steeper slopes. Weighted soil loss as high as 256.73 t ha^{-1} yr^{-1} was observed under >35 per cent slope classes. However, dense forests in all the slope classes registered annual soil loss within the tolerance limit (5 t ha^{-1} yr^{-1}).

Hence, there is an urgent need to adopt suitable soil and water conservation measures to check runoff losses causing loss of soil and nutrients. Integrated watershed management by applying site-specific low cost technology with more emphasis on vegetative measures is required. Steep slopes should be kept under meaningful vegetative cover. Alternative land uses like agrisilvi, agrihorti agroforestry on unstable slopes besides reducing erosion losses generate additional income to farmers and sufficient biomass for nutrient recycling.

The soil conservation measures like bunding, terracing, land leveling, farm forestry, regular monitoring of vegetation status, biotic interference, gap filling, safe disposal of runoff, gully plugging, afforestation of gullies to check erosion and landslides should be taken up immediately. Apart from reducing soil and nutrient losses, reduction in runoff will also help in augmentation of groundwater. Soils rendered unproductive through years of soil and nutrient losses by water erosion need to be made productive by adopting soil and water conservation measures along with integrated nutrient management. The inter-cropping with nitrogen fixing trees and crops is also recommended to improve the productivity status of the eroded areas.

Future Needs

Restoration of ecological balance and prevention of soil erosion, to conserve natural resources are the priority areas. Vegetative measures seems to be the best option as they are cheap, can be applied in large area and are interlinked with socio-economic well being of the village poor. Alternate land use systems like agrisilvi and agrihorti agroforestry need further encouragement in the region. Social forestry plantation should also be encouraged. The capability of these alternate systems to simultaneously produce food, fodder, timber and fiber makes it a suitable system from farmer's perspective and sustainable system from environmental point of view.

References

Bhumbla, D.R. (1976). Soil conservation in Shivaliks. *Indian Farming,* 25(12):3-5.

Dhruvanarayana, V.V. and Ram Babu (1983). Estimation of soil erosion in India. *Journal of Irrigation and Drainage Engineering,* 109(4):419-434.

Dhruvanarayana, V.V., Ram Babu and Venkataramanan, C. (1986). Soil erosion under different agroclimatic conditions in Inida. *Indian Journal of Soil Conservation* 14 (2) 261-263.

Grewal, S.S. (1993). Agroforestry systems for soil and water conservation in Shivalik. In Agroforestry in 2000 AD for semi-arid and arid tropics, NRCAF, Jhansi, India, pp 82-85.

Kale, S.P., A.G. Durgue, M.D. Gund, and R.B. Pawar, (1992). Effect of slope and cropping systems on runoff, soil and nutrient losses and biomass production in shallow Inceptisols. *Indian Journal of Soil Conservation* 20(3) 10-14.

Kumar, S. (1997). Erodibility studies on the soils of mid hill region of Solan district of Himachal Pradesh. M.Sc. theses, Department of Soil Science and Water Management, UHF, Nauni, Solan (HP) India.

Kumar, S, J.C. Sharma and I.P. Sharma (2002). Water retention characteristics and erodibility indices of some soils under different land uses in North-West Himalayas. *Indian Journal of Soil Conservation* 30:29-35.

Kumar, V. (2005). Erodibility status of soils under different land uses in Shivalik foothills of district Solan in Himachal Pradesh". M.Sc. theses, Department of Soil Science and Water Management, UHF, Nauni, Solan (HP) India.

Mathan, K.K. and N. Kannan (1993). Influence of rainfall on nutrient loss in sediment and runoff water in two watersheds under contrasting landscape system. *Journal of the Indian Society of Soil Science* 21(3) 606-608.

Middleton, H.E. (1930). Properties of soils which influence soil erosion. USDA Agri. Tech. Bull. No. 178: 1-16.

Mishra, P.R. and M. Sarin (1987). Sukhomajri-Nada-A new model for ecodevelopment. *Business India,* Nov. 16-19, 1987, pp 78-79.

Mittal, S.P., R.K. Aggarwal and J.S. Samra (2000). Fifty years of research on sustainable resource management in Shivaliks. Central Soil and Water Conservation Research and Training Institute, Research Centre, Chandigarh.

Morgan, R.P.C. (1979). Soil Erosion. Longman Group Limited, London.

Narain, P., Ram Babu and M.S. Ram Mohan Rao (1993). Soil erosion map of West Bengal. . *Indian Journal of Soil Conservation* 21:6-10.

NBSSLUP (2004). Soil resource management reports. National Bureau of Soil Survey and Land Use Planning, Nagpur.

Sharma, I.P., S.Kumar and J.C. Sharma (2000). Rainfall-runoff-soil loss relationships under different landuses in mid-hill region of Himachal Pradesh. *Indian Journal of Soil Conservation* 28:91-97.

Sharma, J.C., J.Prasad, A.R. Bhandari and K.R.Sharma (2003). Prioritisation and management needs of Kawal khad watershed based on erosional soil loss in Southern Himachal Pradesh using remotely sensed data and GIS techniques. *Journal of the Indian Society of Soil Science* 51: 291-297.

Sharma, J.C., J. Prasad and A.R. Bhandari (2002). Effect of watershed characteristics on soil erosion in southern Himachal Pradesh using remote sensing and GIS techniques. In (Eds. Dhyani, SK, Tripathi, KP, Singh, R., Raizada, A., Sharma, N.K., Mishra, A.S., Shrimali, S.S., Dhyani, B.L., Sharma, A.R. and Khola, O.P.S.): Resource Conservation and Watershed Management - Technology Options and Future Strategies. IASWC, Dehradun, India, pp 157-163.

Singh, G., Ram Babu, P. Narain, L.S. Bhushan and I.P. Abrol (1992). Soil erosion rates in India. *Journal of Soil and Water Conservation* 47 (1) :97-99.

Wischmeier, W.H. and D.D. Smith, (1978). Agriculture Handbook No. 587, USDA, Agriculture Research Centre.

Wood, J.C. (1987). Important factors influencing water infiltration and sediment production on arid land in New Mexico. *Journal of Arid Environment* 12(2) 117-118.

Yadav, R.P., R.K. Aggarwal, S.L. Arya, Pratap Singh, Ram Prasad, P. Bhattacharya, A.K. Tiwari and M.K. Yadav (2005). Rainwater harvesting and recycling for sustainable production in small agricultural watershed. Technical Bulletin No. T-50/C-11. Central Soil and Water Conservation Research and Training Institute, Research Centre, Chandigarh.

Zatcher, D. (1982). Soil erosion, Elsevier Scientific Publishing Company, New York.

Chapter-5

Potential Agroforestry Systems for Mid Hill Himalayas

P S THAKUR AND C L THAKUR
Department of Silviculture and Agroforestry
University of Horticulture and Forestry, Nauni-173 230, Solan, (HP)

ABSTRACT

This article is an attempt to describe potential agroforestry systems for mid hills zone of Himalayas. The existing traditional land use systems with separate allocation to agriculture and forest are inadequate not only to fill gaps between demand and supply for food, fuel, fodder and timber but also to check land degradation and soil erosion. The adoption of common agroforestry systems, namely agrisilviculture, agrihorticulture, agrisilvihorticulture, agrihortisilviculture, silvimedicinal, silvipastoral and hortipastoral along with appropriate management practices shall be suitable and viable option for checking soil erosion, improving productivity and economic status of the farming community. The functional components of agroforestry systems, however, differ greatly in size, texture and nature depending on the necessity of rural population. The fuel, fodder, timber and fruit trees dominate in the hilly regions, whereas commercial species like poplar, eucalyptus, shisham, etc constitute the major functional component in lower Shivaliks. Farming communities successfully grow food crops like wheat, maize, soybean, pea, black gram; vegetable crops like tomato, cabbage, capsicum; tuber crops and medicinal crops with multipurpose woody perennials. The findings indicate that agricultural crops can be integrated with woody perennials without any adverse impact on growth, production or physiological status of food crops, except when growing immediately beneath tree canopies. Our experience is that planting timber, fuel-fodder or fruit tree species in combination with agricultural crops and medicinal and aromatic herb species pave the way for diversified products and better economic returns to the farmers. Multipurpose tree species outside the forest have played very important role in catering to the day to day requirements of the rural population, beside environmental conservation. Different farming systems consisting crops + trees have yielded encouraging results and have not only helped diversification but enhanced productivity of land per unit area. Intercropping medicinal and aromatic herb species, especially with fruit and commercial timber further makes agroforestry more profitable.

Introduction

It is a necessity to develop potential, viable, acceptable, diversified and sustainable cropping systems, which ensures increased and sustained crop production with minimum depletion of the critical resources available. Exponential growth in human and livestock population has largely been responsible for causing great imbalances between demand and supply of basic needs. The increased demand for trees for fruit, timber, fuel, fodder and construction in both, rural and urban areas is amenable for much of the deforestation. Today, subsistence kind of traditional cropping sequences appear inadequate to sustained flow of multiple outputs, especially with context to hill farming, which is more difficult and complex because of the topography. Dominance of small and marginal farmers, fragmented land holdings, sloppy and erosion prone cultivable lands, undulating terrain, small villages difficult to approach, scanty irrigation, unawareness, non-inclusion of high value cash crops etc. are some of the constraints, which restrict production processes. A stage has almost reached, where monoculture can no longer sustain the farming economy. Moderate efforts to develop and devise appropriate and viable means to diversify productivity by integrating multipurpose tree species and other economically useful woody perennials with high value cash crops, simultaneously or sequentially on the same piece of land will certainly change production scenario. There are good number of agroforestry systems, which are suitable for the upper Shivalik to fulfill the twin goals of diversification and sustainable production. However, the concerted efforts for further improvement of production efficiency as well as compatibility of system components (woody perennials and agricultural corps) are essential, which needs to be looked into at interdisciplinary level. Today, there is a huge shortfall of basic needs, besides an enormous damage to the resources; renewable as well as non-renewable on account of their abuse and overuse. So, to bridge the gap between supply and demand, efficient management of natural resources is the key issue. The huge area in the upper Shivaliks contains bushes or lesser-known tree species. This is the right time to exercise option to convert this huge low productive and less exploited land into a productive gold mine by adopting suitable agroforestry practices. However, in order to transform present day practices into high tech scientific adventure, there is an urgent need to formulate hypothesis through conductive reasoning, test and validate the hypothesis through deductive experimentation and strive to attain predictable understanding that will have wide applicability. The inclusion of trees on the farmland paves the way for the improvement and diversification of existing systems, but nevertheless creates complex biological interactions, which may not necessarily bring the expected yield advantage. The complex and diverse components of the system interact with one another and influence the micro-climate and soil conditions. The interaction between the functional components is of dynamic nature and may prove competitive, complementary and/or supplementary at any phenological phase of arboreal and arable components. Climate change is bound to influence production of agricultural as well as fruit or forest tree crops. The adverse impact of changing climate needs to be understood clearly in order to develop suitable models combining suitable tree-crop components. The blend of traditional knowledge with scientific knowledge is essential for improvement of technologies. The selection of proper multipurpose tree species for agroforestry, followed by appropriate management approach, are the key factors for improving overall productivity per unit of land. While perusing to achieve such tall targets, the efforts must be made to educate farming communities regarding tangible and intangible benefits of agroforestry.

Why Agroforestry ?

The degradation of upland mountain watersheds at an alarming rate is increasingly becoming a serious problem. It is severely undermining the economic, social and environmental security of large population living in the region. In addition, the exponential growth in human and livestock population over the years coupled with the declining production, increased soil erosion and declining cultivable land have increased gaps between supply and demand for timber, fuel and fodder. The growing appetite for trees for fuel, fodder and construction in both rural and urban areas has largely been responsible for much of the deforestation. The existing land use systems with separate allocation to agriculture and forest are not adequate to meet demands for food, fuel, fodder, timber, paper and pulp. The scenario on fuel, fodder, timber and fruit production in the submontain mid hill region is not satisfactory. There exist imbalances between demand and supply. The deficiency is met from the farmland and other lands outside the forests. Therefore, agroforestry, which is an intersectoral intervention spanning two important and at times competitive land use systems i.e. agriculture and forestry seems to be panacea for overcoming most of the problems related to the alleviation of poverty and socio-economic instability. Agroforestry holds great potential for increased availability of on-farm wood products, energy sources, improved carbon sequestration and diversified products. Short rotation plantations consisting eucalyptus, poplars, shisham, acacias and pines will go a long way in improving socio-economic status of smallholders. Raising silvipastoral agroforestry system combining tree species with improved grasses like *Setaria* and *Napier* for production of quality, nutritious and palatable fodder and forage will greatly increase milk production. Fruit based agroforestry systems combining high value cash crops (HVCC) like medicinal and aromatic herbs, floriculture, vegetable and tuber crops will be highly profitable. Several medicinal and aromatic plant species are shade loving so multipurpose tree species will not only protect herb species from erratic high heat, frost, wind velocity but also provide additional monetary returns in the form of commercial timber or other products. Moreover, the preference in all countries is towards greater proportion of cultivated material. Agroforestry, indeed, is a convenient and strategic option for cultivation of medicinal and aromatic plants beneath canopies of multipurpose tree species (fuel, fodder, timber, fruits, pulp etc.). The integration of trees on the farmland paves the way for the improvement and diversification of existing systems, but creates complex biological interactions (Lawson and Kang 1990; Ong *et al.,* 1991; Khybri *et al.,* 1992; Rao *et al.,* 1998; Gillespie *et al.,* 2000; Thakur and Singh 2002; Thakur and Dutt, 2004). Diversification and sustainability in production are the two main goals and there is an opportunity to explore the possibility of growing commercial tree species with high value cash crops (HVCP) such as medicinal plant species on the farmland without compromising quality of either product. The option of simultaneous growing of crops + trees under well managed agroforestry practice helps to reduce gaps between demand and supply (Thakur and Dutt 2003; Rao *et al.*, 2004; Thakur *et al.*, 2007). The critical resources like solar radiation, water and soil nutrients can be put to optimum use through diversification methods like agroforestry and intercropping (Noguchi and Ichimura 2004; Thakur and Kumar 2006; Thakur and Dutt 2007). Growing of trees on the farmland leads to better economic returns and diversification at the farm level (Yadava and Singh 1996; Thakur *et al.*, 2005; 2007). Trees provide farmers with marketable products, such as timber, fuel wood, fodder, fiber, fruits and medicines, all of which earn extra income. Trees increase ecosystem biodiversity and can help mitigate global climate change by sequestering carbon. The appropriate management practices are useful for minimizing competition between the components of the systems (Cannel 1983; Thakur and Dutt 2003; Thakur and Kumar 2006; Thakur *et al.*, 2007). Different agroforestry systems have been developed and numerous tree crop combinations have been tried by researchers all over the world, with an eye on increasing the economic situation of the farmers (Gillespie *et al.*, 2000; Rao *et al.*, 2004; Thakur and Kumar 2006; Thakur *et al.*, 2007).

Main Agroforestry Systems and Functional Components

The existing agroforestry systems, which are commonly found in the upper Shivalik region, are agrisilviculture, agrihorticulture, agrisilvihorticulture, agrihortisilviculture, silvipastoral, commercial plantation and energy plantation. Tree component mostly include fuel, fodder and timber species. Some of the tree species namely, poplar, eucalypts, *Dalbergia, Toona, Grewia, Morus, Bauhinia, Celtis, Acacias,* Pines, *Robinia, Leucaena, Terminlia chebula* (Harar), *Terminalia bellirica* (Bahera), *Sapindus mukorosi* (Ritha), *Quercus* (Ban oak), *Ficus* spp, bamboo, bushes like *Woodfordia* and *Indigofera* (grasslands) etc. dominate the systems. Agrihorticulture combination predominately includes stone fruits (peach, plum, apricot, almond), pomegranate, wild pomegranate, walnut, apple, pear, lemon, kinnow, mango, cherry, litchi, kiwi, strawberry. The arable functional components include, wheat, maize, mustard, lentil, sunflower, soybean, black gram, green gram; vegetable crops like, tomato, capsicum, pea, cabbage, cauliflower, tuber crops, medicinal herbs (*Aloe vera, Chlorophytum,*

Andrographis, Withania, Valeriana, Ocimum, Digitalis, Stevia), floriculture plants (carnation, gladiolus, roses, chrysanthemum, marigold) etc., which are potential intercrops. Grasses like *Setaria, Napier, Chrysopogon, Themeda* and *Apluda* are potential components of silvipastoral systems in the region. However, the performance, production potential, compatibility, sustainability, profitability and suitability of functional components will vary from system to system and management levels.

Medicinal Agroforestry

The benefits of the presence of tree canopies on the farmland can be harnessed for better production potential of several medicinal and aromatic herb species. The author of this chapter has been able to successfully cultivate more than ten important medicinal and aromatic herbs beneath tree canopies of multipurpose tree species. The findings indicate the positive influence of tree shade on the quality and quantity of herb yield. This is essential and important. Our own findings over the years reveal that there does not seem to be any adverse impact of the trees on the growth and production potential of medicinal and aromatic plants. Timber, fruits, fodder and fuel tree species have been found beneficial for the successful cultivation of medicinal herbs; the yield advantage i.e. the herb yield plus the benefit from arboreal components and / or fruit yield assures greater monetary benefits. The adoption of medicinal agroforestry where tree species with medicinal values in place of non-medicinal tree species can be integrated with medicinal and aromatic herbs for even better economic returns. The establishment of this type of agroforestry systems, especially in the upper Shivaliks with medicinal trees as shade providers will not only ensure higher returns but provides a sustainable source of fresh and quality raw materials to the pharmaceutical and cosmetic industries. The conservation of genetic biodiversity and healthcare will be additional advantages. The buyback arrangement with pharmaceutical firms will further boost large scale cultivation of medicinal and aromatic plants. Smallholders and economically marginalized farming communities can be targeted beneficiaries. This is true that the economic feasibility and quality products will be the main rational for inclusion of Medicinal and Aromatic Plants under agroforestry systems. Besides, the approach will generate several opportunities for rural employment.

Why medicinal herbs?

The tremendous awareness for herbal medicines (phytomedicines) world over has led to the extensive use of these medicines for two reasons, i) because these are less expensive and within the reach of common people who can not afford costly health services and ii) that the side effects are minimum. This trend over the years has sharply increased the demand for raw materials by the multinational pharmaceutical and cosmetic industries in India and abroad. The domestic manufacturers of herbal based medicines have further widened the gap between demand and supply. To meet this demand several thousand tones of medicinal and aromatic herbs are unscientifically extracted from the forest ecosystem in the Himalayas resulting a trade, which runs into millions of dollars every year. In the recent years, there has been a shift from growing traditional food crops towards high value cash crops such as medicinal and aromatic plants, which are in great demands. Medicinal and aromatic plants offer multiple benefits including one of monetary benefits to unemployed people world over. Besides securing the healthcare needs of a large number of people, medicinal plants are the exclusive source of some drugs even for the modern medical treatment. The global importance of medicinal and aromatic plants is evident from a huge volume of trade at national and international levels. The demand for medicinal plant based raw material is increasing roughly by 15-20% annually. Conservative estimate put world trade of medicinal plants at 5 trillion USD by 2050. The natural sources are in no position to keep pace and meet this skewed demand. India is considered treasure house of valuable medicinal and aromatic plant species. Out of approximately 15000 plant species found in India; 3000 are medicinal species. North-West Himalayas in Himachal Pradesh is the natural home for about 1100 species of medicinal and aromatic plants and 350 species of ethno botanical importance. The government of Himachal Pradesh has initiated appreciable step at the right time to make this hilly state an 'Herbal State'. Through National Medicinal Plants Board, India has already prioritized medicinal and aromatic herb species for cultivation under different agroclimatic zones of the country.

Why cultivation of medicinal herbs?

Roughly 80% of the rural population in India and other developing countries of the world, exclusively exploit forest resources for meeting their health requirements. This *per se* is the area of great concern to every one may be environmentalists, planners or governments. Unscientific, unsystematic and uninterrupted extraction / removal (slaughter harvesting) of medicinal and aromatic plants from their natural habitats are collectively contributing to the demise of a number of species. The increasing demand for herbal medicines is threatening to extinct several precious medicinal herb species and can spark global healthcare crisis. This practice has resulted in shrinking bioresource availability year after year. In case the extraction of these herbal species continues unchecked for few more years, which seems to be a real possibility, several precious medicinal and aromatic species may acquire the status of endangered species. The scenario is, indeed, grim. Though India has rich diversity but the growing demand is exerting heavy pressure on the existing resources causing a number of species to slip either to threatened or endangered category. Approximately 230 medicinal and aromatic species in India are threatened. The present practice of use and over exploitation of medicinal species from their natural home to meet huge demand for raw material may push many more herbal species into threatened status. There are no strict policies or rules for judicious extraction of these herbs. Several natural sites (home) of these herbal species in Himalayas have completely been denuded. Sooner or later the commercial cultivation of medicinal and aromatic herbs, which are in greatest demand by the pharmaceutical and cosmetic industries at home and abroad, will have to be taken up in order to bridge the gap between demands and dwindling supplies. The shift from collection to cultivation approach will assure sustained supply of raw material and reduce *in situ* harvesting pressure. The truth is that most of the medicinal and aromatic plant species grow wild under the forest canopies, so are shade tolerant. Agroforestry, which is an alternate land use system, seems to play crucial role, where cultivation of medicinal and aromatic herbs along with multipurpose tree species including trees of medicinal importance will ensure production as well as conservation of herbal biodiversity. Cultivation of medicinal and aromatic plants on the farmland will have several socio-economic impacts on rural unemployed population. Large scale cultivation of medicinal herbs is one option to generate income. For threatened medicinal and aromatic species cultivation *per se* is convenient conservation strategy.

The author and his group have successfully cultivated medicinal and aromatic plant species under the canopies of fuel, fodder, timber and fruit tree species without compromising quality and quantity of herbs. The herb species, which have been grown under agroforestry conditions include, *Digitalis lanata, Matricarria chamomella, Withania somnifera, Valeriana jatamansi, Salvia sclaria, Andrographis paniculata, Tagetes minuta, Ocimum sanctum, O. basilicum,* and *Spilanthes acmella.* The presence of tree canopies have been found beneficial for better growth and production. Integration of tree species did not influence aroma or oil content of herbal species. Slight decline in herb yield was noticed in medicinal and aromatic plants growing immediately beneath tree canopies. This, *per se,* reflects interspecific above and belowground competition for available resources. This is true that the initial investments and cost of cultivation of medicinal and aromatic herbs will certainly be higher in comparison to conventional crops because of the purchase of planting materials like seeds, rhizomes, roots, seedlings, etc. but the cost of cultivation in the subsequent years declines as part of the produce can be used as source materials for future. The performance and production ability of intercrops can be further improved by adopting suitable agronomic practices. The use of chemical fertilizers and organic manures was found beneficial for minimizing competition for critical resources. Different agroforestry systems developed by combining medicinal and aromatic herb species with clones of *Populus deltoides* , planted at 4 x 4 m and 6 x 4 m spacing on the farmland indicate better option for diversification of farming systems. The data obtained for growth, herb yield and physiological parameters of medicinal and aromatic herb species indicated no substantial adverse impact of the presence of tree species. Application of fertilizers and organic manures to the medicinal herbs was found very useful for enhancing growth and production ability as well as photosynthetic efficiency of all the herb species. Nitrogen @100 kg ha^{-1} and FYM @ 4 t ha^{-1} in the presence as well as absence of tree plantation proved most effective. Growth with respect to height, leaf number, spread, flower number, branch number and leaf area index, all indices of better growth, was significantly higher in fertilizer and FYM treated plants in comparison to control plants without fertilizer and FYM doses. Similar was the trend for yield parameters, where production was greater with the use of fertilizers and FYM even in the presence of tree species. The herb yield was more under the systems when applied with different doses of fertilizers in comparison to sole crops. Nitrogen was found most effective. The use of fertilizers was found to reduce competition between the components. Vermicompost @ 3-4 t ha^{-1}substantially improved production ability of medicinal herb species. This is desirable since better growth and production in herbal species means higher herbal yield, thus better economic returns. The findings revealed that appropriate doses of fertilizers and / or organic manures may be used for successful cultivation of important medicinal herb species under rainfed agroforestry systems. Medicinal herbs can be grown successfully for better economic returns under agroforestry systems at least during one growing season in a year. On the basis of the quantum of information generated over the years provide ample reasons to conclude that the cultivation of high value cash crops (HVCC) like medicinal and aromatic herbs with multipurpose tree species, especially commercial timber is a profitable alternative for traditional cropping systems. We have observed that high value cash crops like, medicinal herbs can successfully be cultivated as intercrops with arboreal component without any adverse impact on the performance and production ability of these crops, except when grown immediately beneath tree canopies. The presence of tree species, particularly commercial timber species and fruit trees provide additional monetary benefit. The bio-economic appraisal of medicinal + poplar agroforestry systems has exhibited better benefit: cost ratio over the sole cropping. The economic analysis indicates that medicinal herbs as intercrops with multipurpose tree species, especially timber and fruit tree species provides BCR between 1.5 to 2.4, which can further be improved by adopting suitable agronomic practices and use of organic manures. The use of fertilizers and organic manures can further improve the performance and production ability of intercrops by minimizing competition for critical resources. It is suggested that cultivation of medicinal herbs in combination with multipurpose tree species, especially poplar and temperate fruit species may be adopted for better economic returns from the systems. Intercropping of medicinal and aromatic herb species with commercial timber, fruit trees or medicinal tree species holds great potential.

Management of Functional Components

It is true that the appropriate solar radiation environment under tree canopies is the deciding and driving force for triggering on and off of vital physiological and biochemical processes associated with the performance and production potential of under storey crops. The transmission of photosynthetically active radiation (PAR) through tree canopy can be regulated by adopting suitable canopy management options. Medicinal plants are shade loving and love to grow and satisfactorily complete their phenological phases even at shade intensity up to 75 per cent. Canopy management of multipurpose tree species integrated on the farmland as agroforestry tree with crops, is essential to minimize competition for critical resources. However, some time the removal of aboveground portion of trees results in enormous decrease in photosynthesis and vigour of the trees, which is not desirable. There exists a positive correlation between the amount of intercepted incident solar radiation and total aboveground production of foliage and wood biomass. Thus, it is of utmost importance to have full understanding of the crown architecture and growth in order to design appropriate canopy management strategies. The main purpose of canopy management and tree spacing is to make multipurpose tree species suitable and compatible with the associated arable crops in any agroforestry system. In case of horticultural component, proper knowledge of cultural practices, especially pruning schedule, is essential for better fruit production. Same is true for associated food, medicinal and aromatic herb species, vegetable, tuber and floriculture crops, especially when grown as intercrops. The yield advantage from agroforestry systems depend upon judicious sharing of critical resources among the functional components. Reports are available, where the positive and/or negative impacts of the presence of tree canopies on the performance of under storey vegetation have been indicated. The major management options for manipulating trees in agroforestry are based on the alternation of light (solar radiation) profile and moisture distribution. Tree canopies of agroforestry trees are mainly managed through pruning, coppicing, pollarding, lopping, partial crown removal, appropriate plant spacing and orientation (planting geometry) etc. These options have both advantages and disadvantages. The impacts of such practices can some times have drastic implications as the removal of aboveground portion devoids tree of exposed effective photosynthetic surface, which may result in the drying of tree. In addition, the management of tree canopy is bound to have direct bearing on the root characteristics as well as growth, vigor and biomass production of the tree itself (Kang *et. al* 1981; Basappa, 1986; Dutt and Jamwal, 1987; Heering, 1995; Puri and Gargya, 1995; El-Fadl, 1997, Singh *et. al.* 1998; Thakur, 2000; Thakur and Sehgal, 2001, 2003, 2004; Thakur and Singh 2003; Thakur *et al.* 2005). Author's experience with canopy management of some of the important tree species of temperate region, indicated enormous influence of management options on vigor and biomass production potential. In agroforestry, tree canopy reduces and modifies incident solar radiation intensity available to the under storey crops; this situation will be highly beneficial for supporting cultivation of medicinal plants under agroforestry intervention.

Medicinal Tree Species as Agroforestry Components

Beside commercial timber species as one of the functional components of medicinal and aromatic plant based agroforestry intervention, the scientific plantation of tree species with immense medicinal values on the farmland with medicinal herb species will be highly profitable. The current emphasis on the integration of medicinal tree species on the field is far from sufficient. The potential tree species with medicinal values can be prioritized for cultivation with annual, biennial and perennial medicinal and aromatic plant species for different agroclimatic zones. There is long list of tree species with known medicinal values; some of them commonly found on the farmers field, which do not require any mention here. However, more systematic approach shall be fruitful.

Water: a critical constraint for large scale cultivation of medicinal herbs

Cultivation of medicinal and aromatic herbs, possessing abundant medicinal values, is fast becoming a convenient and attractive approach to partially substitute for conventional agricultural crops. Farmers are taking up cultivation of important medicinal and aromatic herbs mainly to increase economic returns. Cultivation of medicinal and aromatic herb species is considered a viable and lucrative option to enhance economy of farmers. Water, however, is one of the most important critical resources, which will remain a decisive constraint for large scale cultivation of medicinal and aromatic plants in particular and overall performance of agroforestry system in general. Major cultivable land in upper Shivaliks is under rainfed conditions, which may increase further on account of continuously depleting water sources. It is expected that in the future, climate variability will increase leading to more frequent droughts. The consequences would be overall decline in production levels. Erratic and lesser precipitation and abrupt changes in minimum and maximum temperatures are areas of great concern. There is need to critically examine responses of medicinal and aromatic plant species as well as all the functional components of agroforestry systems to water stress and to identify herb species/ crops capable of coping and sustaining growth and production even under water stress conditions. Integrated agronomic approaches to enhance water productivity of functional components of the systems shall be useful.

Climate Change and Agroforestry Systems

Implications of global warming and climate change on flora and fauna including production ability of food crops have been the hot topic of discussion and debate at national and international levels during the last decade. Alarmed by the possible adverse impacts of global and regional climate change on flora and fauna, there has been serious concern world over in understanding vulnerability of agricultural, fruit and tree crops and prioritizing short and long term strategies to reverse the negative effects. The Intergovernmental Panel on Climate Change (IPCC) in its Fourth Assessment Report (AR 4) published in December 2007 has already indicated warming trend during the 20th century and its repercussions on biodiversity and existing flora and fauna. The report gives greater attention to the integration of climate change with sustainable development and the interrelationships between mitigation and adaptation with special mention of regional issues, uncertainty and risks, climate change and water. The climate change in the form of elevated carbon dioxide, drought and extreme temperature will not only result in drastic decline in production potential of annual and perennial crops but may also cause elimination of some of the important germplasms. The senior author in a recently concluded study (Thakur *et al.*, 2008) has for the first time reported that regional climate change exerts significant influence on the phenological phases of multipurpose tree species. The findings indicate that phenological phases namely, leaf emergence, flower initiation phase and growth period of eleven agroforestry tree species advanced during a short period of eight years, which could be the result of climate change at the regional level. The increase in average maximum temperature by 3.7 °C within 8 years (1999-2006) and a sharp decline in annual rainfall from 1213.9 mm during 1999 to 969.4 mm in 2006 probably were deciding factors. There is every possibility that fast changing climate, especially temperature, level of CO_2, moisture unavailability, etc. will have adverse impact on survival and production ability of all categories of plants. Functional components of agroforestry systems including medicinal and aromatic plants, food crops, vegetable, tuber, floriculture crops will be no exception. Uncongenial microsite climate, result of global warming or environmental degradation, may threaten medicinal and aromatic plants at their natural habitats. Several overexploited medicinal and aromatic herb species may be put at 'risk' in the near future. The result can be the bulging list of endangered species. The short and long term strategies to bring wild medicinal and aromatic plant species under cultivation needs serious rethinking on scientific lines. Development of promising cultivation methodologies seems to be the need, both for better production and biodiversity conservation of medicinal and aromatic plants. Climate variability at the regional rather than global level needs to be looked into more closely in order to more precisely understand impacts, especially on medicinal and aromatic herbs.

Research Agenda

Future agroforestry research is bound to have changed vision and strategies in order to make this alternate land use more lucrative and profitable venture. Today we are equipped with adequate technologies to handle present day agroforestry, however, there is every probability that this package may not suffice to meet the challenges and problems in future. We don't have the comprehensive information on the advantages and disadvantages of different canopy management practices on various aspects related to growth, vigor, root characteristics, biomass production, longevity, etc. We have not yet completely understood the intricate resource sharing between different components of the system; extent of adverse impact on the production potential of crops, which undergo drastic change at the advent of canopy management. A great deal of work has been done on various aspects in agroforestry although the underlying physiological processes are yet to be understood completely. There is no specific data to establish any kind of relationship between canopy management practices and carbon sequestration. Quantification of above and belowground biological interactions, especially at tree-crop interface still remains inconclusive. It is important to workout the proportion of woody and arable components, which can be accommodated per unit area, in order to enhance yield advantage and benefits from the agroforestry systems by minimizing competition. Suitable management of tree canopies, permitting adequate transmission of photosynthetically active radiation beneath canopies has been found more critical for improving productivity of associated arable crops. Appropriate tree management practices not only regulate biomass production ability and maintain growth and vigor for extended period but also help associated crops to perform better under agroforestry interventions. Diversification of farming systems by growing short rotation timber, paper, pulp, fuel and fodder tree species with high value cash crops on the farmland is a viable option for increasing income of farmers. So we need to continuously refine strategies and update the technology package based on the past experiences and failures to make agroforestry more viable, beneficial and widely acceptable option. The task is not yet over. More and more wild medicinal and aromatic plant species need to be brought under cultivation. New crop-tree combinations need to be worked out for better compatibility of functional components in the context of the fast changing climate. Upgradation and refinement of cultivation methodologies for functional components is essential for production (profit) as well as sustainability. In brief, priority areas like i) identification of suitable functional components for agroforestry, ii) development of suitable and low-cost cultivation agro techniques, iii) shift from collection to cultivation in case of medicinal and aromatic plants, iv) screening of water stress tolerant medicinal herb species, v) species able to cope with high temperature and CO_2 (tolerate climate change), vi) technology to improve production and utilization, vii) conservation, viii) post harvest handling and ix) strong and organized market linkages, need immediate attention before scope and potential of agroforestry intervention can be realized.

Conclusion

Diversification of farming systems by combining annuals, biennials or perennials; arable and arboreal functional components as intercrops on the farmland is a viable option for increasing income of farmers. Inclusion of commercial timber species, fruit trees, fuel and fodder or medicinal tree species with food, vegetable, tuber, medicinal, floriculture crops and improved grasses will not only encourage diversification but also is great approach to handle environmental degradation. There is tremendous scope for improving socio-economic status of farming community of the region through agroforestry practices.

References

Bassapa, B. (1986). Coppicing of silver oak (*Grevillea robusta*). *My Forest. 22: 1-22*

Cannel, M.G.R. (1983). Plant management in agroforestry: manipulation of trees, population densities and mixtures of trees and herbaceous crops. In: Plant research and agroforestry (P. A. Huxley ed.) Nairobi: ICRAF. Pp 455-487.

Dutt, A.K. and U. Jamwal (1987). Effects of coppicing at different heights on wood production in *Leucaena. Luciana Res. Rep.* 8:27-28.

El-Fadl, M.A. (1997). Management of *Prosopis juliflora* for use in agroforestry systems in Sudan. *Tropical Forestry Report Helsinki.* 16:107.

Gillespie, A.R., S Jose, D.B. Mengel, W.L. Hoover, P.E. Pope, J.R. Street, D.J. Biehle, T. Stall and T.J. Benjamin (2000). Defining competition vectors in a temperate alley cropping system in the Midwestern USA. A Production Physiology. *Agroforestry Systems.* 48: 25-40.

Heering, J.H. (1995). The effect of cutting height and frequency on the forage weed and seed production of six *Sesbania sesban* accessions under irrigated conditions. *Agroforestry Systems.* 30:340-350.

Kang, B.T.; G.F. Wilson and L. Sipkens (1981). Alley cropping with maize and *Luciana* in Southern Nigeria. *Plant and Soil.* 63:165-169.

Khybri, M.L., R.K. Gupta, S. Ram and H.P.S. Tomar (1992). Crop yields of rice and wheat grown in rotation as intercrops with three tree species in the outer hills of western Himalayas. *Agroforestry Systems.* 17: 193-204.

Lawson, T.L. and B.T. Kang (1990). Yield of maize-cowpea in alley cropping system in relation to available light. *Agriculture Forest. Meteorology.* 55: 347-357.

Noguchi, A. and M. Ichimura (2004). Effects of environmental factors on the growth, flowering, and essential oil concentration and composition of sweet basil and spearmint. *Horticulture Research Japan.* 3: 67-70.

Ong, C.K., J.E. Corlett, R.P. Singh, and C.K. Black, (1991). Above and belowground interactions in agroforestry systems. *Forest Ecology and Management.* 45: 45-57.

Puri, D.N. and G.R. Gargya, (1995). Management of *Morus alba* and *Grewia optiva* for degraded lands. *Van Vigyan.* 33:109-113.

Rao, M.R., P.K.R. Nair, and C.K. Ong, (1998). Biophysical interactions in tropical agroforestry systems. *Agroforestry Systems.* 38: 3-50.

Rao, M.R., M.C. Palada, and B.N. Becker, (2004). Medicinal and aromatic plants in agroforestry. *Agroforestry Systems* 60: 107-122.

Singh, C., M.C. Agarwal, N. Kumar, and D.N. Puri, (1998). Biomass production of *Morus alba* under different management practices on degraded bouldry riverbed lands of Doon Valley. *Indian Forester.* 124:252-260.

Thakur, P.S. (2000). Comparative performance of multipurpose tree species growing at the degraded site. *Indian Forester.* 126:895-900.

Thakur, P.S. and V. Dutt, (2003). Performance of wheat as alley crop grown with *Morus alba* hedgerows under rainfed conditions. *Indian Journal of Agroforestry.* 5: 36-44.

Thakur, P.S. and V. Dutt, (2004). Diversification of farming systems by combining commercial tree species with medicinal plants to boost economy of farmers under rainfed conditions. 4th *International Crop Science Congress, Brisbane, Australia,* Pp 1-6.

Thakur P.S. and V. Dutt, (2007). Cultivation of medicinal and aromatic herbs in agroforestry6 for diversification under submontane conditions of western Himalayas. *Indian Journal of Agroforestry. 9:66-76*

Thakur, P.S. and V. Dutt, S. Sehgal and R. Kumar, (2005). Diversification and improving productivity of mountain farming systems through agroforestry practices in northwestern India. *North American Agroforestry Conference, University of Minnesota, Rochester, USA, Abstract No 25.*

Thakur, P.S., V Dutt and A.Thakur, (2008). Impact of inter-annual climate variability on the phenology of eleven multipurpose tree species. *Current Science.* 94:1053-1058.

Thakur, P.S. and R. Kumar, (2006). Growth and production behaviour of medicinal and aromatic herbs grown under hedgerows of *Leucaena* and *Morus*Indian *Journal of Agroforestry* 8:12-20.

Thakur, P.S. and S. Sehgal, (2001). Effect of canopy management on root parameters in agroforestry tree species of temperate region. *Indian Journal Agroforestry.* 2:75-78.

Thakur, P.S. and S. Sehgal, (2003). Growth, leaf gas exchange and production of biomass in coppiced and pollarded agroforestry tree species. *Journal of Tropical Forest Science.* 15: 432-440.

Thakur, P.S. and S. Sehgal, (2004). Influence of canopy management on the performance of multipurpose tree species. *Indian Forester.* 130:639-646.

Thakur P.S., Sehgal S., Dutt, V. and Thakur, C. (2007). Strategies to improve production ability of medicinal and aromatic herbs under rainfed agroforestry systems in northwestern India. Proceedings of International Workshop on Medicinal and Aromatic Plants Held at Chiang Mai University, Thailand from January 15-18.

Thakur, P.S. and S. Singh, (2002). Effect of *Morus alba* canopy management on light transmission and performance of *Phaseolus mungo* and *Pisium sativum* under rainfed agroforestry. *Indian Journal of Agroforestry* 4: 25-29.

Thakur, P.S. and S. Singh, (2003). Influence of canopy size modification of *Morus alba* on productivity of *Vigna mungo* and *Pisum sativum* in agroforestry system. *Indian Journal of Plant Physiology* (special vol.*):* 104-109.

Yadava, A.K. and K. Singh, (1996). Performance of the lemongrass *Cymbopogon flexuosus* under poplar based agroforestry systems. *Journal of Medicinal and Aromatic Plant Science.* 18:290-294.

Chapter-6

Design for Slope Protection by using Bio-Engineering Measures in Himalayan Region

ROHITASHW KUMAR,* VIJAY SHANKARAND MAHESH KUMAR#**

*Department of Soil Science and Water Management, University of Horticulture and Forestry, Nauni, Solan (HP)-173230, India **Department of Civil Engineering, National Institute of Technology, Hamirpur (HP)-177005, India # Department of Civil Engineering, Malviya National Institute of Technology, Jaipur (Raj), India

---

ABSTRACT

Bio-engineering is the use of vegetation, either alone or in conjunction with civil engineering structures, to reduce instability and erosion of slopes. Such structures are advocated in hilly area mainly because they provide one of the best ways to armour slopes against erosion. Bio-engineering, in the context of upland slope protection and erosion reduction, combines mechanical, biological, and ecological concepts to arrest and prevent shallow slope failures and erosion. This chapter presents the design consideration and construction practices for stabilization of slopes using bioengineering technologies along with a brief review on the salient features and characteristics of the bio-engineering processes.

Introduction

Bio-engineering is the use of plant material, living or dead, to alleviate environmental problems such as shallow, rapid landslides and eroding slopes and stream banks. In bioengineering systems, plants are an important structural component. This approach to slope stabilization requires a true partnership among many disciplines, including soil scientists, agroforesters, hydrologists, botanists, engineering geologists, maintenance personnel, civil engineers, and landscape architects. Soil bio-engineering most often mimics nature by using locally available materials and a minimum of heavy equipment, and it can offer roadside managers an inexpensive way to resolve local environmental problems. These techniques can also be used in combination with traditional engineering techniques such as rock or concrete structures. It provides the best way to armour slopes against erosion, and can also provide a significant contribution to soil reinforcement and other anti-failure measures. It is often called "soil bio-engineering" to distinguish it from the bio-medical science that uses the same term. Soil bio-engineering involves integrating living woody and herbaceous materials with organic and inorganic materials to increase the strength and structure of the soil. Howell *et al.* (1991) and Shrestha (1991) have studied the application of bioengineering to the eastern Himalayas. It combines mechanical, biological, and ecological concepts to arrest and prevent shallow slope failures. Biotechnical engineering techniques rely on biological knowledge to build geotechnical and hydraulic structures and to secure unstable slopes. Soil bioengineering combines the use of live plants or cuttings, dead plant material, and inert structural members to produce living, functioning land stabilization systems. This technique uses living plants to control and prevent soil erosion, sedimentation, and shallow slope instability. The bioengineered solution benefits from engineering techniques that use live plant material. Increasing demands on mountain environments to provide resources to meet the requirements of rapidly growing populations (for example, the population of Southeast Asia has doubled in the last 30 years - IUCN, 1992) are placing pressure on the natural systems of surface protection. The removal of forest for the expansion of agricultural land or exploitation of timber resources can significantly alter the hydrology of steep slopes and remove the reinforcing effect which live tree roots can exert on the stability of slopes. Changes in land use have been shown to be correlated with increased incidences of slope failure (Gray, 1978).

Slope stabilization using bio-engineering might be beneficial or detrimental, depending on the local environment and engineering needs of a particular slope. Plants used in combination can provide much greater effects than single plants, e.g. a single grass plant can catch a small amount of debris and reinforce a small volume of soil with its roots. But if a line of grasses is planted across a slope, together they form a continuous chain to catch debris, and can provide a linear rather than a point of reinforcement. The contour line of grass will also increase the infiltration capacity of the soil (Meyer, 1987).

Usefulness and Applicability of Soil Bio-Engineering

Under most circumstances, bio-engineering/ agroforestry can be effectively combined with appropriate and low cost geotechnical applications to provide the most cost-effective, integrated solution to slope stability problems. Soil bio-engineering methods can be cost effective and a useful mitigation solution for site specific problems. These techniques may not effectively mitigate severe bridge scour, severe roadway erosion conditions, or deep seated slope instabilities. It is used for slopes that are at risk of shallow landslides, slumps, sloughing, and surface erosion. Soil bio-engineering is applied to both cut and fill slopes. Alone it is not appropriate for deep-seated landslides, but can be used in conjunction with other engineering methods to treat associated shallow instabilities. This technique can be used to stabilize the slopes of construction sites or to repair disturbed or damaged slopes. Bio-engineering is the most affordable and effective way of enhancing civil engineering structures to increase stability as far as possible. The vegetative structures are also flexible, being capable of absorbing movement and recovering from damage. Bio-engineering is simply part of wise and sustainable asset management since it helps to ensure the life of physical structures, and reduces overall maintenance costs. Plants reduce the supply of debris from degrading slopes. Soil bio-engineering uses particular characteristics of vegetative components and integrates specific characteristics of structures with vegetation. Structures play a critical role in the establishment of vegetation on steep slopes or in areas subject to severe erosion.

Integrated Planning and Requirement

Basic approaches to upland slope protection and erosion control can be divided into two general categories: living and nonliving. The living approach, which uses live plant materials, can be further divided into two specific categories: vegetative plantings and soil bio-engineering. Vegetative plantings are conventional plantings of grasses, forbs, and shrubs used to prevent surface erosion. Soil bio-engineering utilizes live plant parts to provide soil reinforcement and prevent surface erosion. Nonliving approaches use rigid constructions, such as surface armoring, gravity retaining walls, and rock buttresses. Vegetation can be used in conjunction with nonliving structures to create vegetated structures. Soil bioengineering combines biological elements with engineering design principles. The requirements for both must be considered when planning and designing the measures. The soil bioengineering techniques are generally appropriate for immediate protection of slopes against surface erosion, shallow mass wasting, cut and fill slope stabilization, earth embankment protection, and small gully repair treatment. Soil bio-engineering uses particular characteristics of vegetative components and integrates specific characteristics of structures with vegetation. The resulting systems and their components have benefits and limitations that need to be considered prior to selecting them for use. The vegetative components may be herbaceous species or woody species.

Herbaceous species

- Herbaceous vegetation, especially grasses and forbs, offers long-term protection against surface (water and wind) erosion on slopes
- Vegetation helps to prevent surface erosion by
    - Binding and restraining soil particles in place
    - Reducing sediment transport
    - Intercepting raindrops
    - Retarding velocity of runoff
    - Enhancing and maintaining infiltration capacity
    - Minimizing freeze-thaw cycles of soils susceptible to frost

Woody species/Agroforestry trees

- More deeply rooted woody vegetation provides greater protection against shallow mass movement by:
    - Mechanically reinforcing the soil with roots
    - Depleting soil-water through transpiration and interception
    - Buttressing and soil arching action from embedded stems

Structural components

Structures may make it possible to establish plants on slopes steeper than would normally be possible.

- Structures stabilize slopes during the critical time for seed germination and root growth.
- It may be a retaining structure or grade stabilization structure.
- A retaining structure is usually required to protect and stabilize steep slopes.
- Low retaining structures at the toe of a slope make it possible to grade the slope back to a more stable angle that can be successfully revegetated without loss of land at the Crest.
- Grade stabilization structures are used to control and prevent gully erosion.
- A grade stabilization structure reduces the grade above it and dissipates the excess energy of flowing water within the structure itself.
- Debris and sediment tend to be deposited and trapped upstream of the structure. This, in turn, permits establishment of vegetation behind the structure, which further stabilizes the ground.

CHARACTERISTICS OF SOIL BIO- ENGINEERING FOR SHIVALIK HIMALAYAS

The basic principles that apply to conventional soil erosion control also apply in general to soil bio-engineering in shivaliks. These principles are mostly common sense guidelines that involve planning, timing, and minimizing site disturbance as well as the design of individual measures themselves. Applicable principles can be summarized as follows:

Fit the soil bio-engineering system to the site

This means considering site topography, geology, soils, vegetation, and hydrology. Avoid extensive grading and earthwork in critical areas and perform soil tests to determine if vigorous plant growth can be supported. At a minimum, collect the following information:

(i) Topography and exposure

- Degree of slope in stable and unstable areas.
- Presence or lack of moisture
- Observation of existing stable slopes in the vicinity

- Type and density of existing vegetation in areas with and without moisture and on slopes facing different directions. e.g. certain plants grow well on east-facing slopes, but will not survive on south-facing slopes
- Look for areas of vegetation that may be growing more vigorously than other site vegetation. This is generally a good indicator of excess moisture, such as seeps and a perched water table, or it may reflect a change in soils.

(ii) Geology and soils

- Geologic history and types of deposits (colluvium, glacial, alluvium, other)
- Evidence of past sliding
  - Determine whether the slide occurred along a deep or shallow failure surface
  - Leaning or deformed trees may indicate previous slope movement or downhill creep
- Determine soil type and depth

(iii) Hydrology

- Determine the drainage area associated with the problem area
- Determine the annual precipitation
- Calculate peak flows or mean discharge through the project area
- If a seep area was noted, locate the source of the water
- Determine whether the water can be intercepted and diverted away from the slope face

Retain existing vegetation whenever possible

Limit removal of vegetation

- Limit cleared area to the smallest practical size
- Limit duration of disturbance to the shortest practical time
- Remove and store existing woody vegetation that may be used later in the project
- Schedule land clearing during periods of low precipitation whenever possible

Design and Construction Consideration in Shivaliks

For design of soil bio-engineering structure the following considerations has to be taken into account.

Design consideration

Earthwork

- Modify soil structures during construction
- A steep undercut or slumping bank requires grading to flatten the slope for stability.
- The degree of flattening depends on the soil type, hydrologic conditions, geology, and other site factors.

Scheduling and timing

- Select suitable agroforestry plants, methods and construction timing
- Planning and coordination are needed to achieve optimal timing and scheduling
- The seasonal availability of plants or the best time of year to plant them may not coincide with the construction season or with tight construction schedules.

Vegetative damage to inert structures

- Vegetative damage to inert structures may occur when inappropriate species or plant materials that exceed the size of openings in the face of structures are used.
- Vegetative damage does not generally occur from roots.
- Plant roots tend to avoid porous, open-faced retaining structures because of excessive sunlight, moisture deficiencies, and the lack of a growing medium

Moisture requirements and effects

- Excessive amounts of clay, silt, and organic matter are not desirable for structures.
- Free drainage is essential to the mechanical integrity of an earth retaining structure and also important to vegetation

- Establishment and maintenance of vegetation, requires the presence of some fine particles and organic matter in the soil to provide adequate moisture and nutrient retention
- Biological requirements can be satisfied without compromising engineering performance of the structure
- Needs of plants and the requirements of structures must be taken into account when designing a system

Properties of Soil

Density and compaction, permeability, pH and fertility are to be assessed and improved according to the plants chosen for bio-engineering measures.

Erosion risk like soil erodibility, rainfall erosivity, slope (height and angle) and reinforcement with geotextile. Geotechnical properties like shear strength, slope, and factor of safety, selection of suitable soil materials, soil density and moisture are to be evaluated for selecting suitable treatments and reinforcements.

Construction Considerations

- Plant species must be suitable for the intended use and adapted to the site's climate and soil conditions
- Rooted plants and vegetative cuttings are living materials and must be handled properly to avoid excess stress, such as drying or exposure to heat.
- They must be installed in moist soil and be adequately covered.
- The soil must be compacted to eliminate or minimize air pockets around the buried stems.

Design and Construction Techniques

Various soil bioengineering techniques, their applications, effectiveness, and construction guidelines has been presented here. The methods for construction and plantation of trees and structures are different depending upon the type of vegetative material. To maintain stability and prevent further surface erosion and shallow, rapid land sliding, planting "islands" can be created by constructing willow fences with a brush layer base to provide additional slope protection. Once the fence is constructed, the area behind the willow fence can be filled with soil (preferably a silt loam). Shrubs and small trees would be planted within these terraces. To mitigate erosion, a rock apron can be placed at base of the slope. To complement the buttressing effect, live stakes would be installed in the rock apron. The different type of installation methods are as:

Grass Planting

Grasses are effective for stabilizing surface erosion. Grass seed is spread on to the slope, armouring the surface. Alternatively, grass is hand-planted in lines across the slope. The lines armour the slope and catch debris. Angled lines planted by hand may also help to drain the surface, but catch little debris. This helps in reducing slope angle, surface erosion (rills and gullies). This trap sediment, capture and utilize both surface and subsurface water. Slow water movement through sand layer occurs. Root development increases soil strength and slope stability.

Seeding

The most common method of hill and slope stabilization is the seeding of a grass and herb mixture. Various types of seeding methods include:

- *Dry-seeding*: Applied on flat slopes with rough surfaces
- *Cover crop seeding*: On steeper slopes where it is necessary to cover the soil quickly
- *Tree and shrub seeding*: On less steep slopes with rough surfaces and no erosion problems
- *Hydro-seeding*: On steep slopes which have a smooth surface and mild climate and are also mainly in forests
- *Bitumen straw seeding*: On exposed areas and areas mainly above the forest line
- *Jute mesh straw seeding*: On very steep slopes and on failure edges
- *Wire mesh straw seeding*: On areas where rockfall could happen or rocks could break out of the soil.

Live stake

Live staking involves the insertion and tamping of live, rootable vegetative cuttings of multipurpose trees into the ground. A system of stakes creates a living root mat that stabilizes the soil by reinforcing and binding soil particles together and by extracting excess soil moisture. This is an appropriate technique for repair of small earth slips and slumps that are frequently wet. This technique is for relatively uncomplicated site conditions, when construction time is limited, and an inexpensive method is necessary (Fig. 1). It may be used for pegging down surface erosion control materials. This enhances conditions for natural invasion and the establishment of other plants from the surrounding plant community. This can be used to stabilize intervening area between other soil bioengineering techniques, such as live fascines.

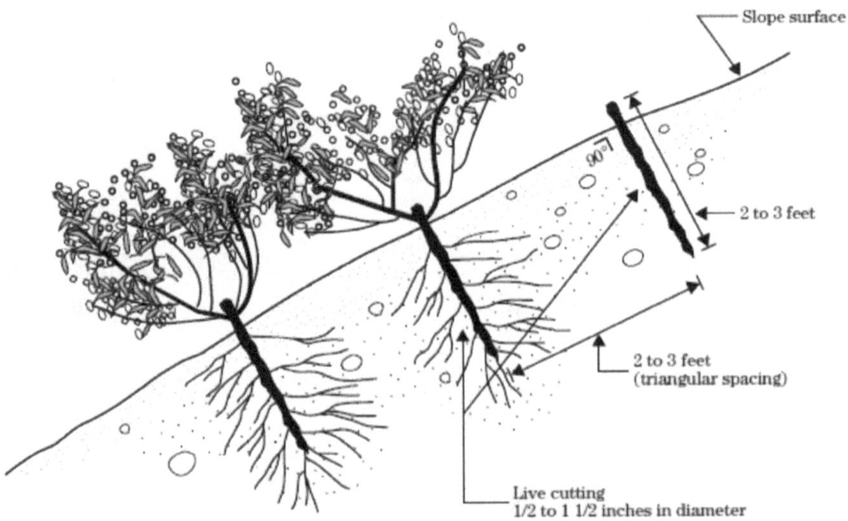

Fig. 1 : Details of Installation of a Live Stake (*Source:* EFH, 1992)

Live fascine

Live fascines are long bundles of branch cuttings, of suitable agroforestry species, bound together with at least 2 mm thick wire into sausage-like structures. When cut from appropriate species and properly installed with live and dead stout stakes, they will root and immediately begin to stabilize slopes. They should be placed in shallow contour trenches on dry slopes and at an angle on wet slopes to reduce erosion and shallow face sliding. The lowest third of the bundle should also contain dead branch material which channels the water unhindered. The tip end of the branches always points to the flow direction. Usually the fascines are 30 to 60 cm thick (Fig. 2) and have no length limit.

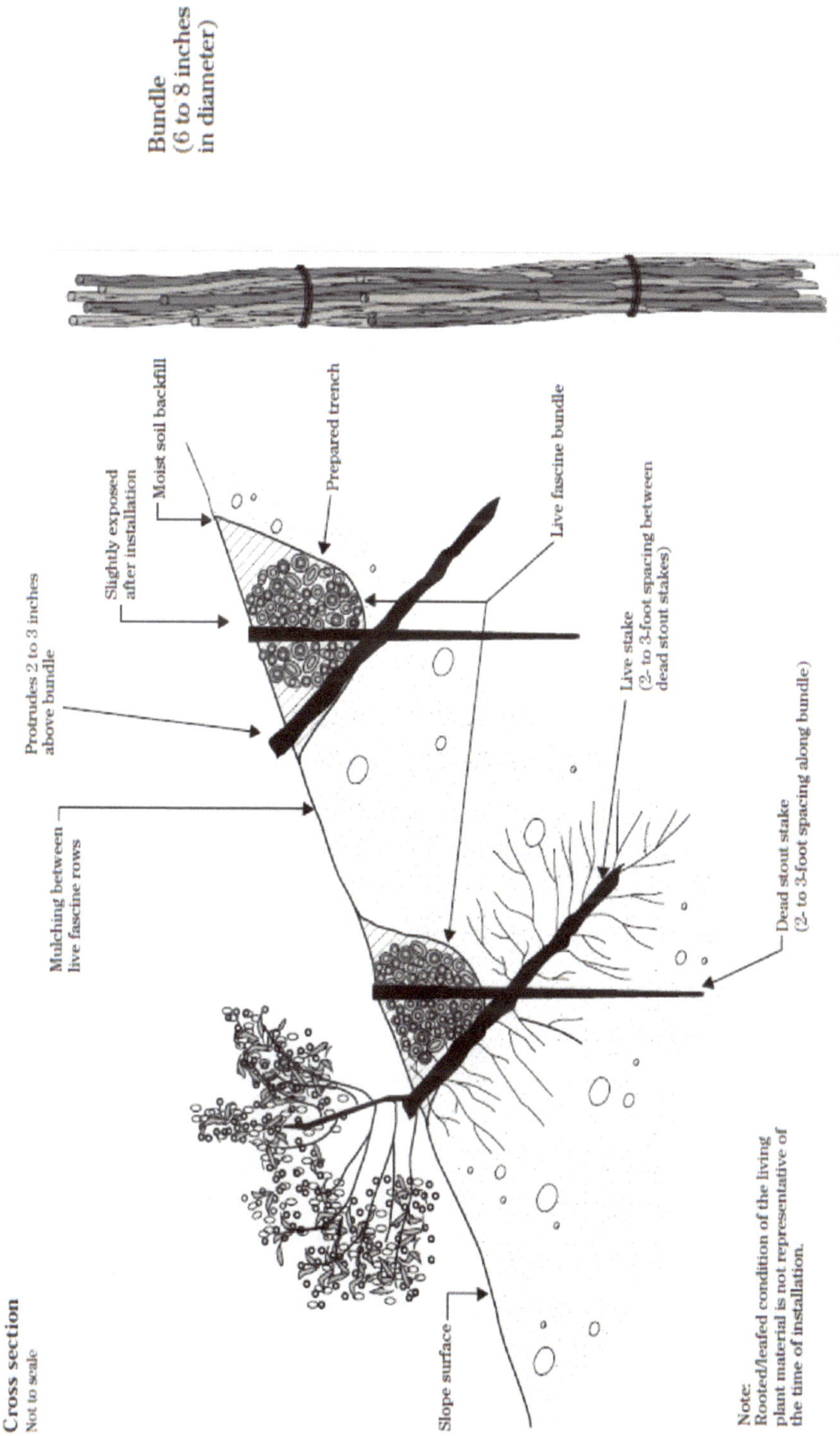

Fig. 2 : Details of Installation of Live Fascines (*Source:* EFH, 1992)

They are put into ditches and then must be covered with soil so that all the branches are embedded and can take root and grow. To prevent the fascines from sliding or washing away, they are fixed every 2 m with wooden poles. This process enhances vegetative establishment by creating a microclimate conducive to plant growth. Little preparation and fast and simple construction is required.

Brush layer

Brush layering is somewhat similar to live fascine systems because both involve the cutting and placement of live branch cuttings on slopes. In brush layering, the cuttings are oriented more or less perpendicular to the slope contour. Brush layering consists of placing live branch cuttings in small benches excavated into the slope. The benches can range from 2 to 3 feet wide. These systems are recommended on slopes up to 2:1 in steepness and not to exceed 15 feet in vertical height. Brush layer branches serve as tensile inclusions or reinforcing units. The portions of the brush that protrude from the slope face assist in retarding runoff and reducing surface erosion. Various types of brush layers are:

- *Willow cuttings*: Have the ability to regrow
- *Hedge brush layer or rooted and unrooted layer*: Combination of willow cuttings and rooted plants, hedge layer or rooted brush layer is a method where only rooted plants are used

It is one of the best methods to stabilize loose rock slopes. Brush layers perform several immediate functions in erosion control, earth reinforcement, and mass stability of slopes such as:

Branch packing

Branch packing consists of alternating layers of live branch cuttings and compacted backfill to repair small localized slumps and holes in slopes (Fig. 3).

Live branch cuttings may range from 1/2 inch to 2 inches in diameter is used for this system. Branch packing are required to be placed parallel to slope contours. It is effective in earth reinforcement and mass stability of small earthen fill sites. These produce a filter barrier, reducing erosion and scouring conditions. This repairs holes in earthen embankments other than dams where water retention is a function and provides immediate soil reinforcement.

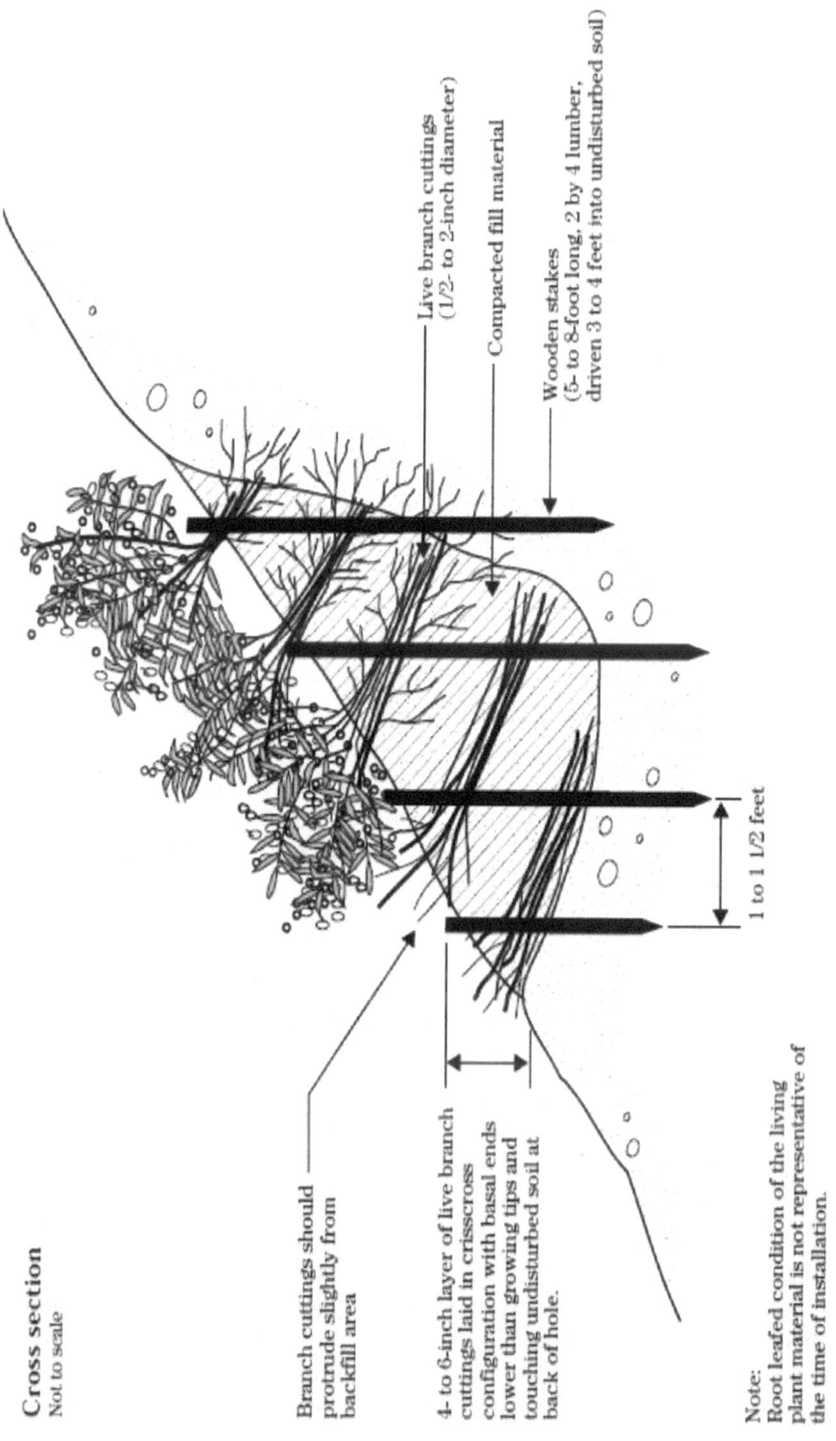

Fig. 3 : Details of Installation of Branch Packings (*Source:* EFH, 1992)

Live gully repair

A live gully repair utilizes alternating layers of live branch cuttings and compacted soil to repair small rills and gullies (Fig. 4). Live branch cuttings may range from 1/2 inch to 2 inches in diameter. They should be long enough to touch the undisturbed soil at the back of the rill or gully and extend slightly from the rebuilt slope face. The installed branches offer immediate reinforcement to the compacted soil and reduce the velocity of concentrated flow of water. This provides a filter barrier that reduces rill and gully erosion. This process is limited to rills or gullies which are a maximum of 2 feet wide, 1 foot deep, and 15 feet long.

Live cribwall

Live wooden cribwall is one of the best method to immediately secure endangered parts of slopes and toes of slopes. A live cribwall consists of a hollow, box-like interlocking arrangement of untreated log or timber. The structure is filled with suitable backfill material and layers of live branch cuttings which root inside the crib structure and extend into the slope (Fig. 5). Once the live cuttings root become established, the subsequent vegetation gradually takes over the structural functions of the wood members. They can be erected as single or double cribwalls and are built from logs and anchor logs held together with nails or bolts. The anchor logs should not be above each other but should be placed alternately. For higher stability against sliding, cribwalls should not be placed horizontally on the slope but at an angle of 10-15 per cent towards the slope. Additionally, the whole construction is secured with 2-2.5 m long iron poles, which are hammered into the ground. The space between the logs is filled with soil material and plants, which should not stick out of the wall more than a quarter of their length. To reach vegetation quickly it is advantageous to use green willow branches and strong rooted pioneer plants. In order to achieve good plant development the face wall should be 50 per cent. This technique is appropriate at the base of a slope where a low wall may be required to stabilize the toe of the slope and reduce its steepness. It is not designed for or intended to resist large, lateral earth stresses. It should be constructed to a maximum of 6 feet in overall height, including the excavation required for a stable foundation. This is useful where space is limited and a more vertical structure is required. This provides immediate protection from erosion, while established vegetation provides long term stability. It should be tilted back or battered if the system is built on a smooth, evenly sloped surface. It may also be constructed in a stair-step fashion, with each successive course of timbers set back 6 to 9 inches toward the slope face from the previously installed course (Engineering Field Handbook (EFH), 1992).

Vegetated rock gabions

Vegetated gabions begin as rectangular containers fabricated from a triple twisted, hexagonal mesh of heavily galvanized steel wire. Empty gabions are placed in position, wired to adjoining gabions, filled with stones and then folded shut and wired at the ends and sides. Live branches are placed on each consecutive layer between the rock filled baskets (Fig. 6). These will take root inside the gabion baskets and in the soil behind the structures. In time the roots consolidate the structure and bind it to the slope. Vegetated gabions are used to stabilize slopes and slope toes where a lot of smaller stones are available. To ensure that the plants are covered with material, it is necessary to lift the mesh and shake it to settle the gravel. The plants should protrude slightly beyond the face of the gabion but not more than 10 cm to prevent desiccation. Finally, the wire mesh is pulled together and sewn shut with wire. Moreover, vegetated gabions are a fast and simple construction to secure wet slopes because they are elastic and improve drainage through plant transpiration. This technique is appropriate at the base of a slope where a low wall may be required to stabilize the toe of the slope and reduce its steepness. Not designed for or intended to resist large, lateral earth stresses. It should be constructed to a maximum of 5 feet in overall height, including the excavation required for a stable foundation. This is useful where space is limited and a more vertical structure is required.

Joint planting

Joint planting or vegetated riprap involves tamping live cuttings of rootable plant material of multipurpose species into soil between the joints or open spaces in rocks that have previously been placed on a slope (Fig. 7).

Alternatively, the cuttings can be tamped into place at the same time that rock is being placed on the slope face. They are used where rock riprap is required. Roots improve drainage by removing soil moisture. Over time, they create a living root mat in the soil base upon which the rock has been placed. The root systems of this mat help to bind or reinforce the soil and to prevent washout of fines between and below the rock units.

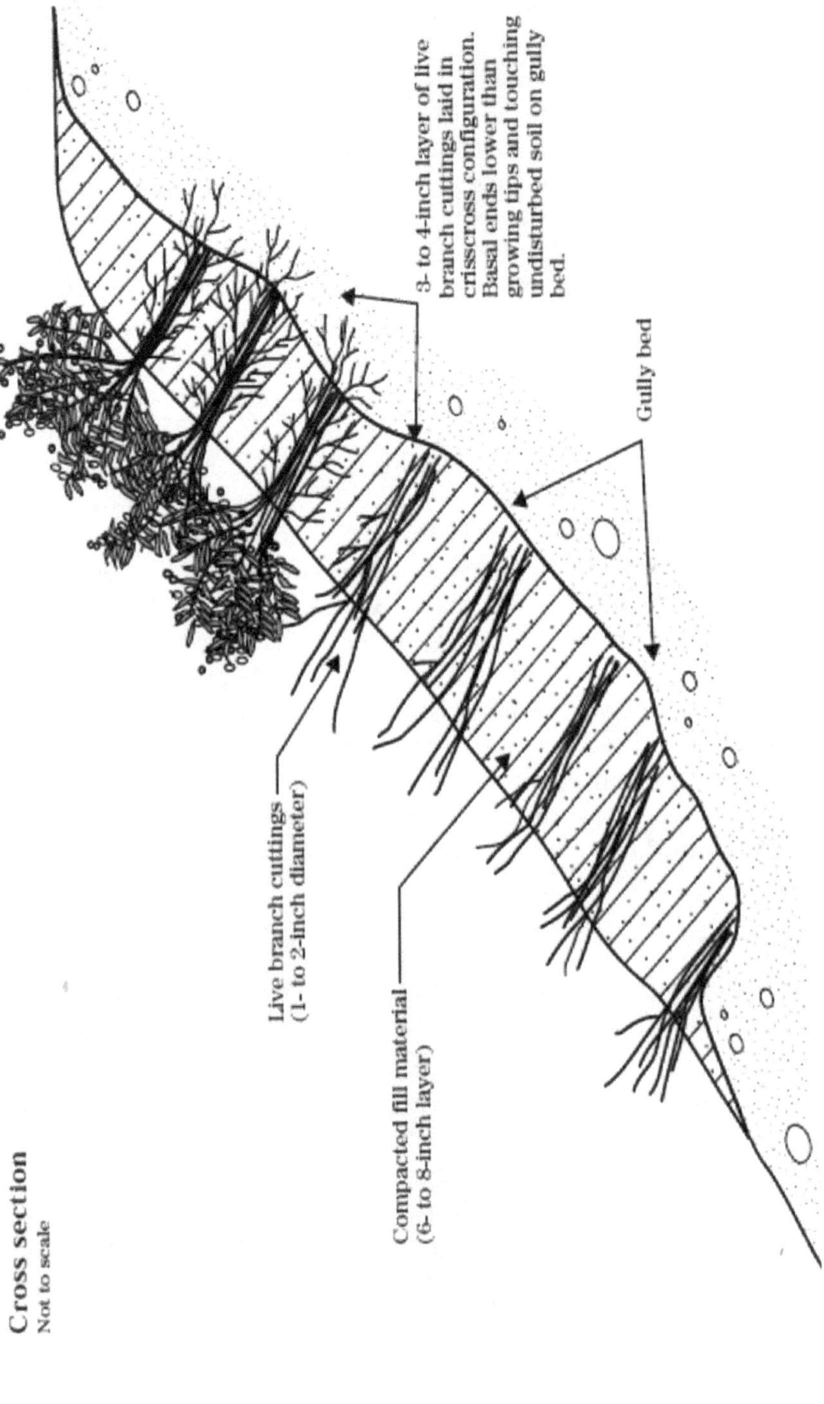

Fig. 4 : Details of Installation of Live gully repair (*Source:* EFH, 1992)

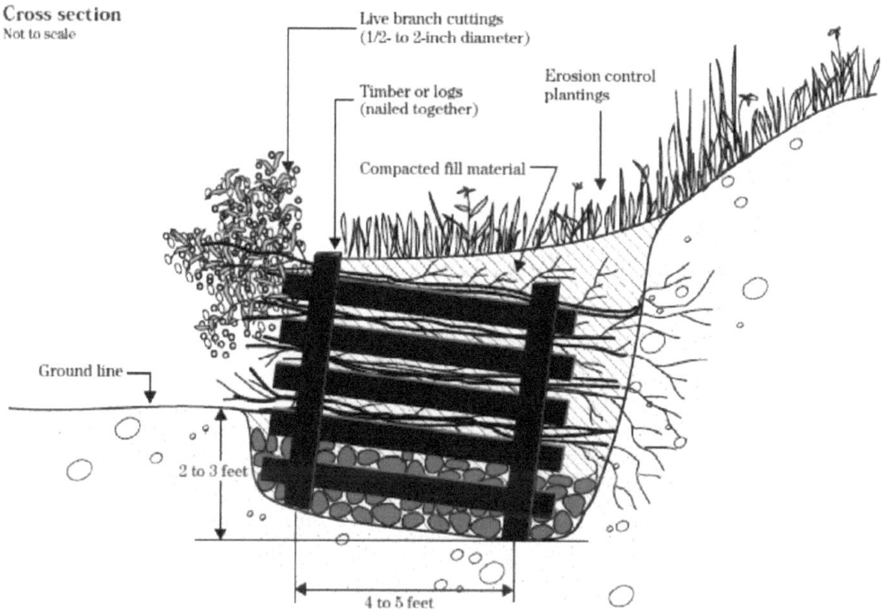

Fig. 5 : Details of Installation of Live cribwall (*Source:* EFH, 1992)

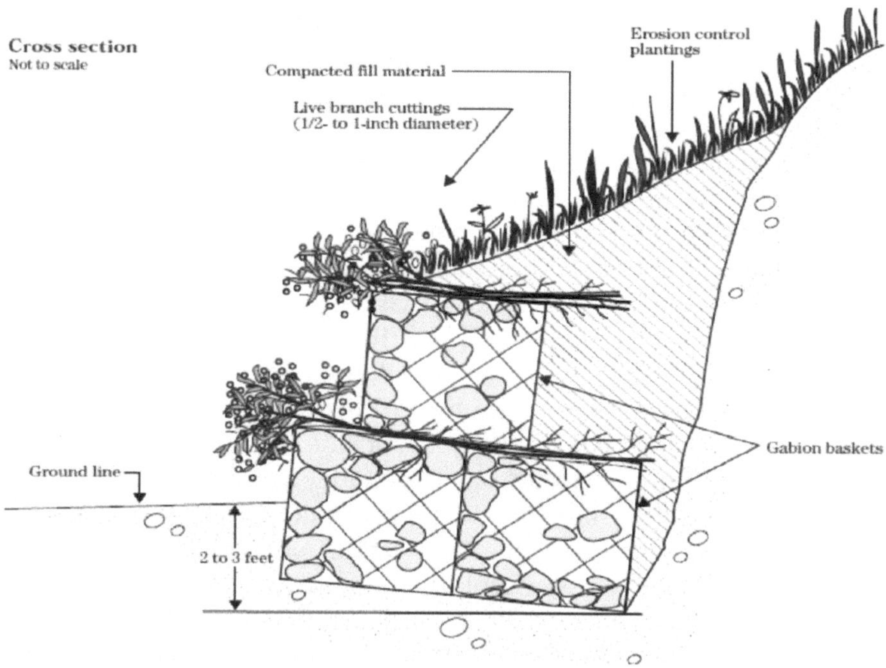

Fig. 6 : Details of Installation of Vegetated rock gabions (*Source:* EFH, 1992)

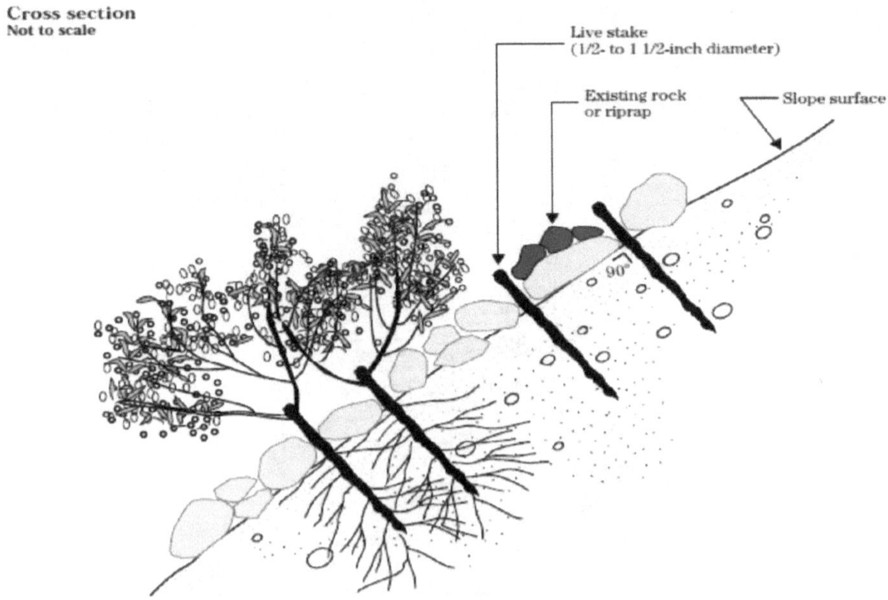

Fig. 7 : Details of Installation of Joint Planting (*Source:* EFH, 1992)

Vegetated rock wall

A vegetated rock wall is a combination of rock and live branch cuttings used to stabilize and protect the toe of steep slopes. Vegetated stone walls are useful for the stabilization of slope toes or steep slope cuts. Vegetated rock walls differ from conventional retaining structures in that they are placed against relatively undisturbed earth and are not intended to resist large lateral earth pressures (Fig. 8). They are flexible, permeable, and durable and can be adapted to every slope angle. During the building of the stone wall, live plants are placed into the joints between the stones so that they reach into the soil behind the stones. The joints must be filled with soil material to ensure plant growth. Green willow branches and rooted plants which have the ability to produce adventitious root systems should be used as vegetation. The branches should not protrude from the wall more than 10 cm to prevent desiccation. Additionally, a greening with hydro seeding is possible. This system is appropriate at the base of a slope where a low wall may be required to stabilize the toe of the slope and reduce its steepness. It is useful where space is limited and natural rock is available. The function of the plants is the stabilization of the construction with their root system and the absorption and transpiration of water by the vegetation, which drains the slope.

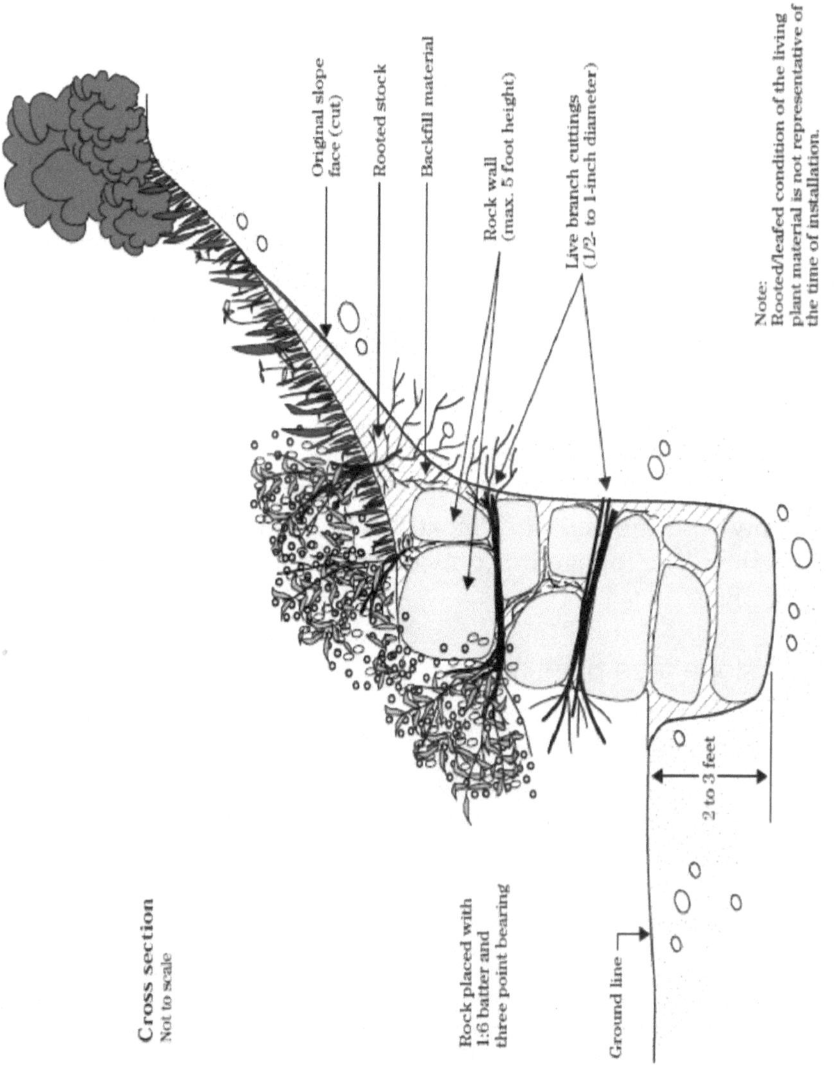

Fig. 8 : Details of Installation of Vegetated rock wall (*Source:* EFH, 1992)

Composite systems

A range of composite systems are commonly used. Examples are:

- *Live check dams*: Armour and reinforce gully beds and catch debris;
- *Vegetated stone pitching*: Provides strong armouring for ephemeral water courses;
- *Planted geotextiles*: Geotextile provides armouring, later supplemented by the vegetation, which also reinforces the soil.

Vegetated structures

Vegetated structures consist of either low walls or revetments (concrete or rock and mortar) at the foot of a slope with planting of fodder, fuel and timber yielding species on the interposed benches. A structure at the foot of a slope protects the slope against undermining or scouring and provides a slight buttressing effect. In the case of low walls, it allows re-grading of the slope face to a more stable angle without excessive retreat at the crest. Vegetation planted on the crest of the wall and the face of the slope protects against erosion and shallow sloughing. In the case of tiered structures, the roots of woody plants grow into the soil and backfill within the structure, binding them together. The foliage in front covers the structure and enhances its appearance.

Low wall/slope face plantings

A low retaining structure at the foot of a slope makes it possible to flatten the slope and establish vegetation. Vegetation on the face of the slope protects against both surface erosion and shallow face sliding. Gravity wall that resists lateral earth pressures by its weight or mass is commonly used (Fig. 9).

Tiered wall/bench plantings

An alternative to a low wall with face planting is a tiered retaining wall system. This alternative effectively allows vegetation to be planted on slopes that would otherwise be too steep. Shrubs and trees planted on the benches screen the structure behind and lend a more natural appearance while their roots permeate and protect the benches. Virtually any type of retaining structure can be used in a tiered wall system. A tiered wall system provides numerous opportunities for adding vegetative values on steep slopes and embankments.

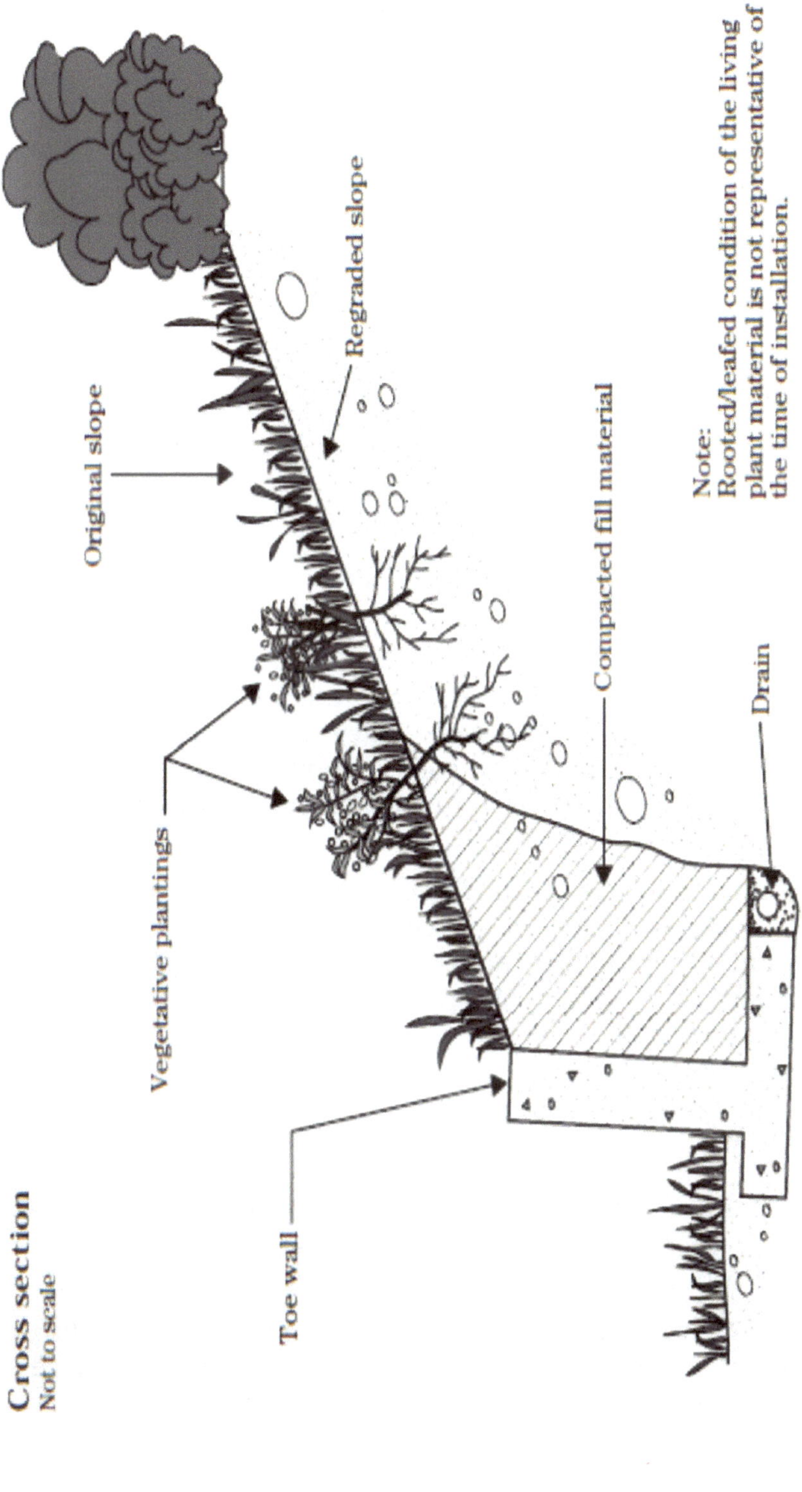

Fig. 9 : Details of Installation of Vegetated rock wall (*Source:* EFH, 1992)

Cribwalls with plantings

A cribwall is a structure formed by joining a number of cells together and filling them with soil, gravel, or rock to furnish strength and weight. In crib structures, the members are essentially assembled in "log cabin" fashion. The frontal, horizontal members are termed stretchers; the lateral members, headers. The frontal spaces between the stretchers in conventional cribwalls provide openings through which vegetative cuttings can be inserted and established in the crib fill.

Quality Control and Drainage

Quality control required at preconstruction stage are:

- Select plant species for conformance to requirements.
- Locate and secure source sites for harvesting live cuttings or commercial procurement.
- Define construction work area limits.
- Fence sites requiring special protection
- Complete and inspection of layout, excavation systems, bench size, shape, angle, preparation of site (i.e., clearing, grading, and shaping), disposal of excess gravel, soil, and debris, depth of excavation, vegetation to be removed/ preserved, stockpiling of suitable soil and/or rock.

Quality control required during construction are inspection of each system component, at every stage, for angle of placement and orientation of the live cuttings, backfill material/rock and stone material, fertilizer, method and quantity applied, lime, method and quantity applied, preparation of trenches or benches in cut and fill slopes, staking, pruning, stock handling and preparation, soil compaction and watering. Also one has to ensure that proper maintenance occurs during and after installation. Inspection for daily for quality control is required. Checking of all cuttings; removal of unacceptable material and use fresh stock for replacement installations is also required. Timing and planting medium to be checked properly and system is required to be provided periodic maintenance.

Conventional grading, contour grading and landform grading is provided in the soil bio-engineering structures for proper drainage of water, which prevent washout of the installed plants and provides longevity to the structure. Conventional graded slopes are characterized by linear planar slope surfaces with unvarying gradients and angular slope intersection. Slope drainage devises are usually constructed in a rectilinear configuration in exposed position. In contour grading slopes are curvilinear rather than linear with unvarying gradients. Slope drainage devices are usually constructed in geometric configuration and in an exposed position to the slope face. Landform grading replicates irregular shapes of natural, stable slopes. It is a characterized by a series of concave and convex forms intercepted with swales and berm combinations to conceal the drains from view. Slope drainage devices either follow natural slope drop line or are tucked away in special swale and berm combinations to conceal the drains from view.

Conclusion

Bio-engineering is effective and economical either alone or in conjunction with civil engineering structures, to reduce instability and erosion on slopes. It provides the best way to armour slopes against erosion, and can also provide a significant contribution to soil reinforcement and other anti-failure measures. A proper design approach and site investigation is required before installation of vegetation on the site. Biotechnical engineering does not replace traditional hydraulic and geotechnical engineering (e.g. geotextiles, or concrete blocks), but complements and improves other technical engineering methods. In order to achieve high levels of strength, physical structures are always required and bio-engineering measures can only complement the civil engineering solutions adopted. Vegetation/ agroforestry species can provide protection and reinforcement of backfill and surrounding slope areas, protection from scour and the undercutting of the foundations and sides of structures and a flexible extension to a wall through large bamboos, shrubs or trees close to it adding to the engineering functions of catching, supporting and buttressing. Properly designed and installed structures help to stabilize a slope against shallow mass movement and protect the slope against rill and gully formation. Soil bio-engineering has unique attributes, but is not appropriate for all sites and situations. Topography, hydrology, climatic conditions, properties of soil, availability of local vegetation and requirement of civil structures should be properly investigated before installation. Depending upon the slope grading and soil characteristics suitable technology can be adopted. A combined process will be more effective for slope stabilization with combination of mechanical and biological (agroforestry) measures. Thus bioengineering measures in Shivaliks will be more effective in controlling the loose strata and hence soil erosion unlike using either vegetative or mechanical measures in isolation.

References

Engineering Field Handbook (EFH) (1992). United States Department of Agriculture, Soil Bioengineering for Upland Slope Protection and Erosion Reduction, 1992.

Gray, D. H. (1978). Role of woody vegetation in reinforcing soils and stabilising slopes. In: Proc, Symp.

Howell, J.H., J.E.Clark, C.J. Lawrance and I. Sunwar (1991). Vegetation Structures for Stabilising Highway Slopes. UK/Nepal Eastern Region Interim Project, Kathmandu.

IUCN (1992). Regional Reviews. IVth World Congress on National Parks and Protected Areas, Caracas.

Meyer, W.P. (ed) (1987). Vegetative soil conservation measures: A Field Manual, Vol. 1. Tinau Watershed Project, Tansen.

Shrestha, V.P. (1991). Biotechnical slope stabilisation in mountain roads. MSc Thesis, University Strathclyde.

Chapter-7

Suitable Agroforestry Systems for Lower Shivaliks of Panchkula, Haryana

PANKAJ PANWAR
Central Soil and Water Conservation Research and Training Institute Research Centre, Chandigarh-160 019

ABSTRACT

The paper presents the results of agroforestry trials that had been conducted in lower shivaliks for assessing soil and water conservation under different land uses. The experiments reveal that agroforestry is a profitable venture compared to monocroping in lower shivaliks, where agriculture is rainfed.

Introduction

The lower Shivalik is composed of an alternating sequence of well indurated, fine grained, purple and grayish-purple sandstones and purple and maroon clays in the lower part and grayish-purple, medium to fine grained standstones in the upper part. The formation attains a thickness of 1,300 m to 1,900 m in the Western Himalaya. The important places of lower Shivaliks are along Sambha-Udhampur Road, Jammu - Srinagar national highway, Nahan, Kalka, Nalagarh, Bilaspur, Sarkaghat regions of Himachal Pradesh. In Kumaon Himalaya, lower Shivaliks are in Kotdwar, Rathwadhab, Mohand, Haldwani and Tanakpur regions (Chaudhri, 2000). The mean size of the sediments of lower Shivalik sediments corresponds to medium to very fine sand (Wentworth, 1922). Clay mineral assemblage of the region includes kaolinite, illite, chlorite, vermiculite, montmorillonite and mixed layers. Illite is the most dominating clay mineral in lower Shivalik formation.

Owing to lack of irrigation facilities, erratic distribution of *Kharif* season rainfall ranging from 490 mm to 1577 mm with coefficient of variation of 28 per cent (Agnihotri *et al.*, 1989), undulating topography and small size of land holdings, the traditional farming has failed to prove to be a profitable venture for the Shivalik foothill farmers. It has been suggested at different forum that an alternative landuse systems be developed to provide a sustained income to the farmers practicing agriculture at high risk. In this venture, Central Soil and Water Conservation Research and Training Institute and other institutes located in Shivaliks had been carrying out extensive work on alternative landuse systems. The paper clubs some of these land uses which had been developed for lower Shivaliks.

Climate and Soil

The area falls in sub-tropical climatic zone with extremes of temperatures in summer and winter. The temperature has a wide range of 5.6°C to 31°C in winter and 15.8°C to 40.4°C in summer on weekly average basis. The rainfall is bimodal in nature and 80 per cent is received between June to September.

The soil belongs to mixed hyper-thermic family of course loamy typic *ustochrepts*, normal to mild alkaline in reaction and loamy sand to sandy loam in texture. The soils are well drained with rapid permeability.

Agroforestry Systems

Aonla (*Embilica officianlis*) based silvipastoral system was established in 1991-92. The study consisted of four treatments *viz.* sole *Embilica officianlis* (no grass), *Embilica officianlis* + *Chrysopogon fulvus* (Dholu grass), *Embilica officianlis* + *Pennisetum purpureum* (Hybrid napier var. NB-21) and *Embilica officianlis* + *Cajanus cajan* (perennial pigeon pea var. ICP-8094). Plot size was 18x12 m and the experiment had three replications. The runoff on an average, over a period of 10 years, was 8.0, 13.1 and 18.6 per cent for aonla + *Chrysopogon fulvus,* aonla + hybrid napier and aonla + perennial pigeon pea, respectively over control (pure aonla). The runoff increased during post-bearing years. The effectiveness of different intercrops in reducing soil loss was in the order of *Chrysopogon* > napier> pigeon pea with 81, 56 and 25 per cent reduction over sole aonla. Contrary to runoff, sediment losses were less during post-bearing stages. Table 1 depicts the yield of above ground biomass of grasses under aonla during various years (Yadav *et al.*, 2005).

Table 1 : Yield of above ground biomass of grasses under aonla

Year	Dry weight of grasses (q ha^{-1})		
	Dholu*	Napier*	Pigeon pea**
1992-93	-	50.00	-
1993-94	50.93	345.10	15.27, 17.99, 2.64
1994-95	65.20	242.10	18.23, 20.98, 7.88
1995-96	57.10	198.20	16.03, 84.08, 1.06
1996-97	52.50	136.30	Nil
1997-98	38.50	128.60	Nil
1998-99	31.90	125.90	Nil
1999-00	50.25	118.60	Nil
2000-01	28.70	124.20	Nil
2001-02	19.81	121.78	Nil
2002-03	11.55	118.64	Nil
2003-04	10.17	113.43	Nil

* Yield of grass fodder; ** Yield of fodder, wood and grain respectively.
Source: Yadav *et al.*, 2005

Grewal, 1993 studied different agroforestry systems to find out their capabilities in reducing soil loss and runoff. It was found that minimum soil loss was 0.07 t ha^{-1} and runoff 0.05 per cent in *Eucalyptus* - Bhabhar grass silvipastoral system, followed by *Acacia catechu* + forage grass landuse system. Maximum soil loss was 5.65 t ha^{-1} and runoff 23 per cent in cultivated fallow (Table 2).

Leucaena is a multipurpose tree from which fodder yield can be obtained through coppicing, pollarding or loping and which allows other annual crops to grow in conjugation with it. Studies have shown that total fodder yield from annual fodder crops plus *Leucaena* (planted at 4 x1 m and maintained at 30 cm height) was 20 per cent more than annual crops alone (Table 3) (Sur, 1994). Along with increased fodder yield integration of *Leucaena* as alley crop improved the fodder quality owing to its high protein content and also covers the risk of crop failure in case of extreme drought conditions.

Table 2 : Soil loss and runoff from different landuse systems in lower Shivaliks.

Landuse system	No. of years of observations	Soil loss (t ha^{-1})	Runoff (%)
Eucalyptus – Bhabhar grass	6	0.07	0.05
Acacia catechu – forage grass	3	0.24	2.00
Leucaena – Pennisetum purpureum grass	3	0.28	4.40
Teak – Leucaena – Bhabbar	3	0.43	3.30
Eucalyptus – Leucaena – Curcuma longa	5	0.59	2.60
Poplar – *Leucaena* – Bhabhar	5	1.54	4.80
Sesamum – *Brassica campestris*	3	2.69	20.50
Cultivated fallow	3	5.65	23.00

Source: Grewal, 1993

Table 3 : Green fodder yield (q ha^{-1}) from alley cropping of *Leucaena* with fodder crops

Crop	Alley cropping				Pure cropping		
	Kharif	*Rabi*	*Leucaena*	Total	*Kharif*	*Rabi*	Total
Bajra – Oat	226	137	126	489	272	170	442
Guar –Barley	142	119	127	388	152	158	310
Sorghum – Wheat	115	116	130	361	136	155	291
Mean	161	124	128	413	187	161	348

Source: Sur, 1994

Thind and Dhillon, 1994 studied the performance of *Rabi* and *Kharif* crops under *Leucaena* and *Eucalyptus* and compared it with control. In one treatment *Leucaena* was lopped at 1.5 m and in other it was kept unlopped. The relative yield of *Rabi* crops obtained under different treatment is given in table 4. It was found that *Rabi* crops performed better when they were grown in association with *Leucaena* which was lopped.

Table 4 : Performance of various *Rabi* crops intercropped with *Leucaena* and *Eucalyptus*.

Treatment	Relative yield (%) over sole crop			
	Wheat	Mustard	Lentil	Oat fodder (green)
Leucaena 1.5 m	109	148	49	132
Leucaena unlopped	16	29	6	43
Eucalyptus	76	61	45	82
Control (without trees)	100 (29.0)*	100 (1.36)	100 (6.68)	100 (337)

* Figures in parenthesis are yields in q ha^{-1} in control.
Source: Thind and Dhillon, 1994

A *Leucaena* based agrisilviculture system with intercrops like maize, black gram and cluster bean was tried. Green fodder yield of *Leucaena* decreased significantly by inter-cropping maize and cluster bean when compared to pure crop of *Leucaena*. At close spacing *Leucaena* gave significantly higher mean yield of air dry fuel wood (9.4 t ha^{-1}) over the other systems. Yield of maize and cluster bean decreased significantly when raised as intercrops in between rows of *Leucaena* at wide spacing (25 cm x 375 cm). The net returns worked out on the basis of the average market prices indicated *Leucaena* at wider spacing grown along with cluster bean gave highest returns. Net returns were seen to accrue from the *Leucaena* with closer spacing at second in order (Table 5) (Mittal and Singh, 1989).

Six Acacia species were planted at a uniform spacing of 3 x 3 m in a silvipastoral system in foot hills of Shivaliks. *Elaliopsis binata* (Bhabhar) grass was planted at a spacing of 75 cm x 75 cm in the month of July. After 10 years of study, in 1992, the species were cut and total yield of wood was obtained for all the species. The highest stem wood obtained was of *Acacia nilotica* (39.03 t ha^{-1}) and least of *Acacia suma* (14.64 t ha^{-1}). Mean air dry weight of bhabhar grass was maximum under *Acacia Senegal* (3.50 t ha^{-1}) and minimum 1.45 t ha^{-1} under *Acacia nilotica* (Table 6) (Agnihotri, 1996).

Table 5 : Mean net return and yield of *Leucaena* and grain crops in intercrop and pure system.

Treatment	Mean yield of *Leucaena* (t ha^{-1})		Grain yield of intercrops (kgha^{-1})	Net Returns (Rs. ha^{-1})
	Green fodder	Air dry fuelwood		
Leucaena (25 x 75 cm)	34.0	9.4	-	3128
Leucaena (25 x 375 cm)	18.9	6.3	-	2360
Leucaena (25 x 375 cm) + Maize	15.9	5.1	1714	3273
Leucaena (25 x 375 cm) + Black gram	18.7	6.0	271	3125
Leucaena (25 x 375 cm) + Cluster bean	14.8	5.0	943	3540
Maize	-	-	2296	2537
Black gram	-	-	403	1559
Cluster bean	-	-	1223	3073
LSD (P=0.05)	1.9	1.3	-	597

Source: Mittal and Singh, 1989

Table 6 : Yield of stem and grass under silvipastoral system

Species	Survival	Air dry yield of tree (t ha^{-1})		Mean air dry Bhabhar yield (t ha^{-1})
		Stem wood	Branch wood	
Acacia seneghal	77	25.42	7.59	3.50
Acacia tortilis	68	33.73	14.20	3.07
Acacia modesta	73	20.24	6.47	2.65
Acacia suma	13	14.64	1.96	1.94
Acacia lenticularis	37	19.11	4.67	2.28
Acacia nilotica	74	39.03	11.49	1.45

Source: Agnihotri, 1996

An agrisilvihorticulture system integrating *Leucaena leucocephala* var k-8 as border row for fuelwood and fodder/organic mulch and manure, *Citrus aurantifolia* var. Baramasi in the next peripheral row, *Carica papaya* var. Honey Dew (1.8 x 1.8 m) in the main plot and shade loving, *Curcuma longa* as an intercrop was established. The system was maintained under irrigated conditions for first five years and latter on under rainfed conditions. *Leucaena* attained an average basal girth of 26.3 and 41.2 cm and provided fuelwood yield of 5.24 and 7.53 t ha^{-1} after the first and second rotation of four years each. The mean airdry branch wood yield from *Leucaena* was 4.70 and 5.11 and foliage 3.1 and 3.2 t ha^{-1} whereas, fruit yield obtained from papaya was 4.24 and 1.94 t ha^{-1} respectively. Lemon gave a yield of 1.10 and 0.86 t ha-1 and turmeric (tuber) 0.89 and 0.53 t ha^{-1} yr^{-1} with and without irrigation, respectively. The mean annual net returns were Rs. 17,746 and Rs. 15,092 under the agroforestry system as against Rs. 7,752 and Rs. 3,342 per ha. from agricultural cropping system (maize + blackgram/Wheat + mustard/maize + wheat) under irrigated and rainfed conditions, respectively (Dyal *et al.*, 1996).

Leucaena based agrisilviculture system with cluster bean gave maximum net returns of Rs. 3,540 per ha^{-1}/yr^{-1} on class II land. *Eucalyptus tereticornis* + *Eulaliopsis binata* system established in class III land gave mean air dry grass yield of 4.2 (used for paper pulp) from October cut and 1.19 t ha^{-1} yr^{-1} (used for fodder) from June cut. The net returns from grass alone were Rs. 4,672 against Rs. 1,679 per ha^{-1} yr^{-1} from rainfed field crops raised on an adjoining plot (Grewal *et al.*, 1992).

Economic appraisal of two silvipastoral systems viz. *Leucaena leucocephala* (var k-8) (L) + *Leucaena leucocephala* (var. EL Salvador) (L) (L L system) and *Leucaena leucocephala* (Var k-8) (L) + Napier grass (N) (L N system) along with traditional crops of Sesamum (S) + Taramira (T) (S T system) was made. L L system was found most economical as B/C ratio ratio at 12 % discount rate was found to be 1.69 against 1.3 from L N system and 1.1 from ST system (Agnihotri *et al.*, 1993).

Conclusion

Agroforestry systems are the best proposition for lower Shivalik. These landuse systems not only conserve soil and water but also provide fodder, fuel, timber, fruit and food. For soil and water conservation objectives silvipasture system are advocated and where objectives are of income generation along with resource conservation agrisilviculture, agrihorticulture or agrisilvihorticulture systems are advocated for lower Shivaliks of Panchkula, Haryana.

References

Agnihotri, Y, R.C. Bansal, Prahlad Singh and S.P. Mittal (1989). Climatological features of Chandigarh, Bulletin T-20/C-1, CSWCRTI, RC, Chandigarh.

Agnihotri, Y. (1996). Economics of silvicultural and silvipastoral systems in Shivaliks foot hills, Annual Report (1996), CSWCRTI, Research Centre, Chandigarh: pp 33-44.

Agnihotri, Y; S.S. Grewal and Shiv Charan Singh (1993). Economical appraisal of silvipastoral systems *vis a vis* traditional agriculture in Shivalik foothills. *Indian J. For.*, 16(2): 138-143.

Chaudhri, R.S. (2000) Geology of the Shivalik group of western and central Himalaya. In Fifty years of research on sustainable resource management in Shivaliks (Mittal, S.P, R.K. Aggarwal and J.S. Samra (eds.) CSWCRTI, Dehra Dun : 3-16.

Dyal, S.K.N, S.S. Grewal and Shiv Charan Singh (1996). An agri-silvi-horticultural system to optimize production and cash returns for Shivalik foothills. *Indian J. Soil Conservation*. 24(2): 150-155.

Grewal, S.S, S.P. Mittal, Dyal Surjit and Y. Agnihotri (1992). Agroforestry systems for soil and water conservation and sustainable production from foothill areas of north India. *Agroforestry Systems*,17: 183 – 191.

Grewal, S.S. (1993). Agroforestry systems for soil and water conservation in Shivalik. In Agroforestry in 2000 A.D for semi arid and arid tropics, NRCAF, Jhansi, India: 82 – 85.

Mittal, S.P. and Pratap Singh (1989). Intercropping field crops between rows of *Leucaena leucocephala* under rainfed conditions in Northern India. *Agroforestry Systems*, 8: 165-172.

Sur, H.S. (1994). Ecological rehabilitation and production improvement in foot hill areas of Shivaliks – concepts, procedures and lessons learnt. Paper presented at "National Seminar on Sustainable Development: Ecological and Social Dimensions" held at Chandigarh, India: Feb. 23-25.

Thind, H.S. and M.S. Dhillon (1994). Comparative effect of *Leucaena* and *Eucalyptus* on soil fertility and productivity of light textured wastelands. Paper presented at "National Conference on Soil and Water Conservation for Sustaining Production and Panchayat Raj" held at Chandigarh, India: Jan. 28-30.

Wentworth, C.K (1922). A scale of grade and class terms for clastic sediments. *Jour. Geol.*, 30: 377-392.

Yadav, R.P., Ram Prasad, R.K. Aggarwal, Y. Agnihotri and J.S. Samra (2005). Aonla (*Embilica officinalis*) based silvi-pastoral systems for soil and water conservation in degraded lands of Shivalik foothills. *Indian J. Soil Cons.*, 33(3): 207-212.

Chapter-8

Agroforestry Systems for Soil Fertility Build up in Degraded Lands of Shivaliks

SHARMISTHA PAL
Central Soil and Water conservation Research and Training Institute Research Centre, Madhya Marg, Chandigarh-160 019

ABSTRACT

Trees play a crucial role in maintaining and regenerating soil fertility through the action of their roots and litter. Trees can also improve soil structure, water holding capacity, crop rooting volume and biological activity in soil by providing biomass. Quick growing species, especially of nitrogen fixing trees when combined with herbaceous crops enhance fertility status of the soil through litter contribution and nitrogen fixation. Nitrogen fixing trees and shrubs, grown in agroforestry systems, are capable of fixing about 50-100 Kg N $ha^{-1}yr^{-1}$. Use of agroforestry is also a viable option for enhancing soil organic carbon density. Soil organic carbon sequestration potential in soils of tropical forest ecosystems varies from 100 to 500 Kg C ha^{-1} yr^{-1}. Improving soil organic carbon pool through agroforestry improves soil quality, biomass productivity and mitigates global warming by reducing the atmospheric concentration of greenhouse gases. Nutrient cycling is another way in which trees draw a large portion of nutrients from deeper soil layer and return it to the soil through crown crash, stem flow and litter fall. In a nutshell, for improving the soil quality, ecological balance and sustainability of agro eco systems , agroforestry is one of the best viable alternative land use system.

Introduction

The relevance of agroforestry systems is many fold higher in hill agriculture as this practice of land use has been found ecologically and economically viable (Bhatt *et al.*, 2001; Bhatt, 2003). Shivalik region is a part of lower Himalayas extending from $30^0 10'$ to $33^0 37'$ N latitides and $73^0 37''$ to $77^0 39''$ E longitudes, covering large portion of Himachal Pradesh, Uttarakhand, Jammu and Kashmir, Punjab and Haryana states in north west India (Kukal *et al.*, 1991). In view of the geological and physical conditions in the Shivalik foothills, silvipastoral systems are most suitable for hilly tracts (Thind and Dhillon, 1994a).

The scattered trees of babul (*Acacia nilotica*) and shisham (*Dalbergia sissoo*) are integral part of the agricultural systems. The Babbar (*Eulaliopsis binnata*) grass which is used in paper industry is one of the major component of pastoral system (Rao, 1987). Nitrogen fixing fodder trees have great potential to grow in this region and this can also provide green fodder specially during lean period (Thind, 2000). The most important nitrogen fixing trees suitable for submountainous zones are *Leucaena leucocephala, Albizzia lebbek, Acacia nilotica* and *Dalbergia sissoo*.

The fundamental assumption in agroforestry, that the integration of trees into farming systems and landscapes can increase soil fertility, productivity and sustainability is based on the observation that soils under forest vegetation generally remain fertile. Agroforestry practices have been shown to influence chemical, physical and biological components of soil fertility. Trees can improve the nutrient balance of a site both by reducing unproductive nutrient losses due to erosion and leaching and by increasing nutrient input through nitrogen fixation. The long tap root system of tree provide anchorage and widely spreading lateral roots bind the soil particles and prevent erosion. Tree cover also provide important habitat for a diverse group of microorganisms. The presence of mycorrhizal roots in most of the trees help them to grow successfully on moisture starved and nutrient deficient soil.

Principle of Soil Fertility Improvement in Agroforestry

Plantation of compatible and desirable species of woody perennials on farmlands can result in improvement in soil fertility. There are several possible mechanisms of soil fertility improvement in agroforestry, which include:

(i) Increase in organic matter content of soil through addition of leaf litter and other plant parts.

(ii) More efficient nutrient cycling within system and consequently more efficient utilization of nutrients that are either inherently present in soil or externally applied.

(iii) Biological nitrogen fixation and solubilization of relatively unavailable nutrients e.g. phosphate through activity of mycorrhiza and phosphate solubilizing bacteria.

(iv) Increase in plant-cycling fraction of nutrients, with their resultant reduction beyond the nutrient-absorbing zone of soil.

(v) Complementary interaction between component species of system, resulting in a more efficient sharing of nutrient resources among components.

(vi) Enhanced nutrient economy, because of different nutrient absorbing-zones of root system of component species

(vii) Moderating effect of additional soil organic matter on extreme soil reactions and consequently improved patterns of nutrient release ability.

Improvement in organic matter status of soil can result in an increased activity of favourable microorganisms in root zone. In addition to nutrient relations, such microorganisms may also produce growth promoting substances through desirable interaction and cause commensalistic effects on the growth of the plant species. Inclusion of trees and woody perennials on farm land can, in the long run, result in marked improvement in physical conditions of soil. The influence of trees on hydrological characteristics can extend from microsite to farm and regional levels. There are evidences that the hydrological characterstics of soil is also favourably influenced by the presence of trees.

Nitrogen fixing trees for soil fertility improvement

Biological nitrogen fixation (BNF) is vital to the terrestrial nitrogen (N) budget, balancing N losses from denitrification and providing N for growth and maintenance of the organisms. Terrestrial ecosystems gain an estimated 130-170 million metric tons of nitrogen (N) annually from biological nitrogen fixation (BNF) (Galloway *et al.*, 1995), with about 40 million metric tons, or 23-31 percent of the total, attributed to forest ecosystems (Burns and Hardy, 1975). At the forest stand level, high rates of BNF are most often reported for actinorhizal and leguminous plants which fix N in symbiosis with soil prokaryotes. For example, *Alnus rubra* (red alder) and *Casuarina equisetifolia*, can fix 130 and 12-85 kg N ha^{-1} yr^{-1}, respectively (Binkley, 1981; Diem and Dommergues, 1990). The possible amount of N fixation by various N fixing trees have been shown in Table 1.

Table 1 : Nitrogen fixation by leguminous tree species

Tree species	N_2 fixed (kg ha^{-1} yr^{-1})
Acacia albida	20
Acacia mearnsii (High land tropics)	200
Acacia mollisima (Arid zone)	200
Casuarina equisetifolia (Humid tropics)	218
Casuarina equisetifolia (Arid zone)	58
Erythrina poeppigiana	60
Gliricidia sepium	13
Inga jinicuil	35-40
Leucaena leucocephala (Arid zone)	110
Leucaena leucocephala (Humid zone)	500
Leucaena leucocephala (Hedge row intercropping)	75-120
Prosopis tamarago	200
Prosopis sp (Arid zone)	30

Source : Nair,1984; Avery, 1987; Singh *et al.*, 2000

Nitrogen fixing trees (NFTs) are often deep rooted, which allows them to gain access to nutrients in subsoil layers. Their constant leaf drop nourishes soil life, which in turn can support more plant life. Aside from enhancing overall fertility by accumulating nitrogen and other nutrients, NFTs establish readily, grow rapidly, and regrow easily from pruning. They are perfectly suited to jump-start organic matter production on a site, creating an abundant source of nutrient-rich mulch for other plants. Many fast-growing NFTs can be cut back regularly over several years for mulch production. The potential nitrogen fixing shrubs can be cultivated as a hedge row on terrace risers, contour bunds or contour trenches (Kadirvel *et al.*, 2003). Woody legume trees, which have potential to fix significant quantities of nitrogen can have a possible influence on yield in the nitrogen deficient soils. Some nitrogen fixing trees have proved to be very efficient in phosphorus uptake and can effectively serve as nutrient pumps for this and other mineral nutrients (Balkrishan and Toky , 1993).

Agroforestry for soil organic carbon build up

A comparison of soil organic carbon (SOC) contents under natural forests and different cultivation practices in soil from different agro climatic region (Table 2) showed lower SOC in cultivated soils as compared to the soils under natural forests. Soil organic carbon, C turnover rate and microbial biomass carbon were many times more in natural forest as well as in mixed forests as compared to arable crop land (Swarup *et al.* 2000).

Table 2 : Carbon pools of under different land management practices

Parameters	Land management systems				
	Sal forest	Mixed forest	Wheat fallow	Pearl millet - mustard - sunflower	Soybean - wheat - fallow
Soil org. C (g m^{-2})	2854	2530	1063	1104	1144
Carbon input (g C m^{-2} yr^{-1})	250	250	281	365	265
Turnover (yr^{-1})					
Soil microbial biomass C (g m^{-2})	87.0	90.0	42.3	31.7	50.8
C input/soil microbial biomass C	2.87	2.78	6.64	11.51	15.74

Source: Swarup *et al.*, 2000

Agroforestry systems have very high potential of carbon sequestration because of increase in organic carbon storage, both in soil and above ground biomass (IPCC, 2000). Indian wastelands estimated to be around 30 million hectare, have the capacity to store additional 24.3 Pg C through afforestration (Gupta and Rao, 1994). Agroforestry is one of the most important carbon sequestration strategy because of carbon storage potential in its multiple plant species and soil as well as its applicability in agricultural lands and in reforestation. According to the estimate, one hectare of new forest is expected to sequester about 6.2 t of carbon annually (Brown *et al.*, 1993). As in other land-use systems, the extent of C sequestered depends on amounts of C in standing biomass, recalcitrant C remaining in soil, and C sequestered in wood products. Proper design and management of agroforestry practices can make them effective carbon sinks. Average carbon storage by agroforestry practices has been estimated as 9, 21, 50, and 63 Mg C ha^{-1} in semiarid, subhumid, humid, and temperate regions. For smallholder agroforestry systems in tropics, potential C sequestration rates range from 1.5 to 3.5 Mg C ha^{-1} yr^{-1}. Agroforestry can also have an indirect effect on C sequestration by decreasing pressure on natural forests, which are the largest sink of terrestrial C. Taking into account the potential for reduction of forest clearance by 5 million ha per year as a result of wide spread agroforestry, lowering of carbon emission from forest burning may be to the level of 1.1 to 1.6 billion t yr^{-1} (Batjes, 1996). The Biomass production of MPTs introduced in natural grassland at 8years of age have been shown in Table 3.

Table 3 : Biomass production of MPTs introduced in natural grassland at 8 years of age

Tree species	Dry biomass production (t ha^{-1})			
	Bole	Branches	Leaves	Total
Azadirachta indica	3.33	2.01	0.51	5.85
Albizia lebbek	13.98	5.52	3.18	22.68
Acacia tortilis	8.46	5.67	0.39	14.52
Anogeissus pendula	1.11	0.45	0.12	1.68
Dichrostachys cinerea	2.64	1.98	0.24	4.86
Pongamia pinnata	4.59	2.52	0.63	7.74
Leucaena leucocephala	9.03	4.53	1.62	15.18
Terminalia belerica	1.89	0.60	0.18	2.67
Zizyphus mauritiana	1.35	0.54	0.09	0.98

LSD (P=0.05)
Source: Rai (1999)

An evaluation of efficiency of three prominent tree species in Shivalik region in terms of carbon storage (Table 4) revealed highest carbon sequestration in *Eucalyptus hybrid* plantation (503.2 t ha^{-1}). In case of *Eucalyptus hybrid, the* carbon sequestration rate was 14.4 t ha^{-1} yr^{-1}, while for *Acacia nilotica* and *Acacia catechu* the rates were 3.6 and 2.1 t ha^{-1} yr^{-1}, respectively.

Table 4 : Carbon storage in three different plantations

Parameters	*Eucalyptus hybrid*	*Acacia catechu*	*Acacia nilotica*
Plantation age (years) (A)	35	35	15
Above ground biomass carbon (t ha^{-1}) (a)	409.9	18.3	21.5
Soil carbon (t ha^{-1}) (b)	35.3	50.3	28.5
Litter biomass carbon (t ha^{-1}) (c)	9.5	0.7	0.1
Root biomass carbon (t ha^{-1}) (d)	45.3	2.7	3.2
Herbaceous / shrubs plant biomass carbon (t ha^{-1}) (e)	3.2	0.4	1.2
Total carbon (t ha^{-1}) (f) = a+b+c+d+e	503.2	72.4	54.5
Carbon sequestration (t ha^{-1} yr^{-1}) (f/A)	14.4	2.1	3.6

Source: Bhattacharya *et al.*, 2009

Soil Fertility Build up Under Different Agroforestry Systems

The mechanisms by which the trees improve nutrient efficiency of soil include nutrient uptake from lower soil horizons, reduction of leaching loss by root systems and balanced nutrient supply. For a tree leaf biomass production (as litter) of 4 tonnes dry matter per ha per year , the potential nutrient return is 80-120 kg nitrogen, 8-12 kg phosphorus , 40-120 kg potassium and 20-60 kg calcium (Palaniappan *et al.,* 1994).

Research conducted in a loamy sand soil of Shivalik region to study the changes in available phosphorus and potash after three years of growth of *Leucaena* and *Eucalyptus* (Figure 1 a and b) revealed that the growing of *Leucaena* for three years considerably improved soil fertility. Surface soil layer (0-15cm) under Eucalyptus and leucaena showed a higher available phosphorus (30.5 and 23.2 kg ha^{-1}, respectively) as compared to control (15.2 kg ha^{-1}). The available potassium in surface layer was found to be 225, 300 and 412 kg ha^{-1}, under control, Leucaena and Eucalyptus, respectively. The available phosphorus and potassium in surface layer was more under eucalyptus, but under leucaena, the build up was up to deeper horizons of the soil profile. The improvement in soil fertility also improved crop yield (Thind and Dhillon, 1994 b).

Nutrient balance study in degraded lands of Shivalik successfully demonstrated a positive nutrient balance in soil with forest plantation comprising various species of *Acacia* (Grewal *et al.,* 1996). The total nutrient input in soil was found to be 173.5, 25.2 and 227.8 kg ha^{-1}of nitrogen, phosphorus and potassium, respectively (Table 5).

Table 5 : Nutrient balance under *Acacia* species plus bhabhar grass system in Shivalik (1974-92)

Particulars of input and output	Biomass (t ha^{-1})	Mean nutrient content (%) N	P	K	Nutrient accumulation (kg ha^{-1}) N	P	K
Input							
Branches and twigs	1.39	0.92	0.14	1.26	12.8	1.9	17.5
Leaf mould	4.12	1.88	023	1.97	76.6	9.5	81.2
Humus	1.96	2.97	2.28	3.08	58.2	5.5	60.4
Soil enrichment	--	--	--	--	25.9	5.3	68.7
Total input	7.47	--	--	--	173.5	25.2	227.8
Output (Bhabhar)	19.80	0.73	0.11	1.13	144.5	21.8	223.7
Net gain	--	--	--	--	29.0	0.4	4.1

Source: Grewal *et al.*, 1996

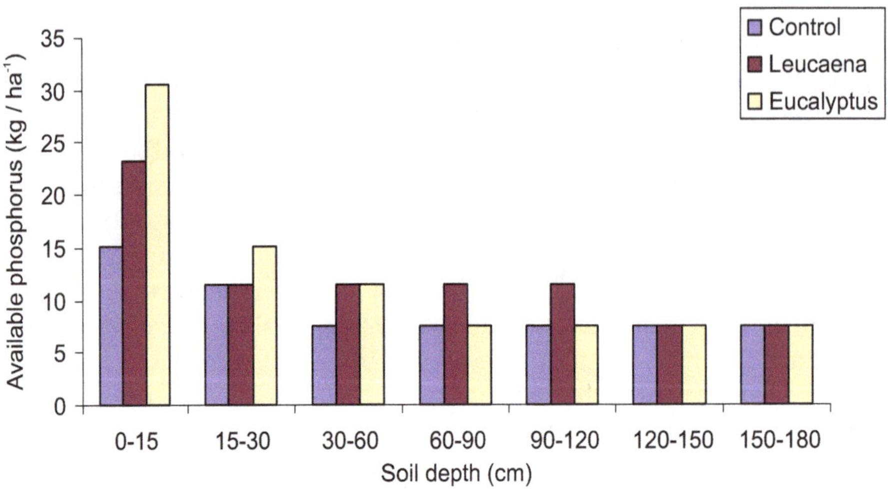

Fig. 1a : Changes in available phosphorus after three years of growth of leucaena and eucalyptus

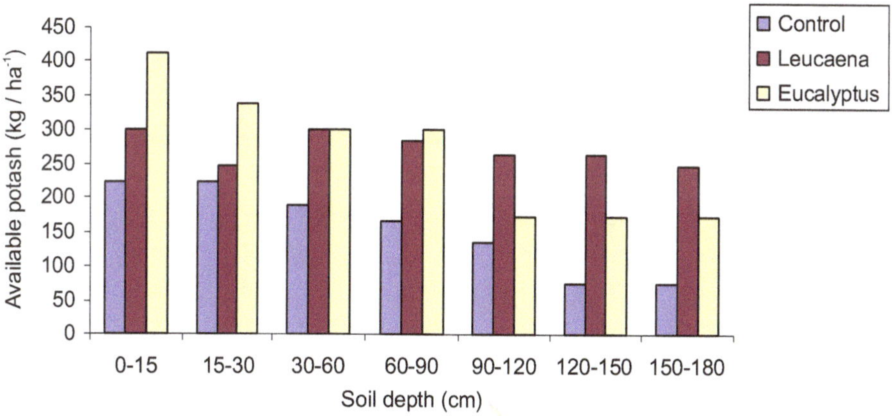

Fig. 1b : Changes in available potash after three years of growth of leucaena and eucalyptus

An increase in soil organic carbon status of surface soil under *Acacia nilotica* + *Saccharum munja* (0.39 % to 0.52%) and *Acacia nilotica* + *Eulaliopsis binnata* (0.44 to 0.55%) have been reported by Samra and Singh(2000). Research in Shivalik region also suggested that *Acacia nilotica* + *Eulaliopsis binata* are conservative, more productive and less competitive with trees and suitable for eco friendly conservation and rehabilitation of degraded lands of Shivalik foothills of subtropical northern India.

Several studies indicated that agro forestry system are more conservative than traditional crops on eroded marginal soil and hence suggested for inclusion in the basket of conservation technologies. A study was conducted to compare the production and conservation potential of two agroforestry systems (*Leucaena - Leucaena* and *Leucaena*-napier grass) with traditional rainfed crop sequence (sesame – rapeseed) and cultivated fallow. The soil profile of sesame followed by rape seed indicated a net deficit of 15, 2 and 14 kg ha^{-1} N, P and K but a gain of 38, 10 and 20 kg and 66, 26 and 57 kgha^{-1} of available N, P and K in *Leucaena-leucaena* and *Leucaena*-Napier, respectively (Grewal , 1996).

Study conducted by Viswanatham *et al* (1999) to document the biomass production of trees and grasses in a silvopasture system on marginal lands of Doon Valley of north-west India concluded that growing of tree species of *Bauhinia purpurea* and *Albizia lebbek* was beneficial for increasing biomass production in the gravely riverbed marginal lands of Doon Valley. The overall performance of *Leucaena leucocephala* was poorest, despite showing the highest survival under these conditions. In a similar study, it was also reported that a silvopasture system involving *Eulaliopsis binata* in association with *Bauhinia purpurea* or *Grewia optiva* appeared to be the most promising for sustained biomass production from the degraded bouldery riverbed lands of Doon Valley (Samra *et al.*, 1999).

In a study carried out to see the effect of perennial mulches on moisture conservation and soil-building properties it was revealed that the order of decomposition of mulch material was *Pueraria hirsuta* > *Broussonatia paprifera* > *Leucaena leucocephala* > *Eucalyptus* hybrid > *Shorea robusta*. Actual assay of soil revealed slight improvement in respect of total nitrogen in case of *L. leucocephala* mulch and overall enhanced soil fertility buildup was with *L. leucocephala* Tomar *et al.* (1992).

A comparison of nutrient conservation under different aonla based silvipastoral system revealed that *Chrysopogon* grass proved very effective in conserving nutrients for longer period due to its compatibility with aonla (Yadav *et al.*, 2005). Besides the optimum production of fuel-wood, fodder and small timber the soil ameliorative property of vegetation on degraded barren watershed is a suitable biological method of soil and water conservation at low cost technology. Mishra *et al* (2000) compared soil properties under high density plantation of three multipurpose trees i.e. *Eucalyptus tereticornis, Melia azedarach* and *Leucaena leucocephala* at Horticultural Research Station at Dhaulakuan, Himachal Pradesh . Soil characteristics like pH, EC, Organic Carbon, available N,P,K, exchangeable Ca and Mg are estimated to assess the soil ameliorative properties. It was noticed that while pH and EC depressed, the status of Organic carbon, exchangeable Calcium and Magnesium, and available N, P, K improved under plantation. Study carried out for the physico-chemical characterization and fertility status of some forest nursery soils of district Sirmour in Himachal Pradesh also revealed a favourable effect of forest plantation on soil pH, CEC and organic carbon, available N,P and K, available Ca and Mg and SO_4-S etc (Raina and Kumar 2000). Similar reports of improved soil quality under different land use systems have been shown elsewhere (Table 6).

Table 6 : Total (%) and available nutrients (kg / ha^{-1}) content of soils under different land use systems

Land use systems	Total N	Available N	Total P	Available P	Total K	Available K	Ex Ca	Ex Mg
Fallow Land	0.021	19.7	186	8.9	1812	116.2	1.92	1.10
Agriculture land	0.039	138	223	12	2281	139.2	2.82	1.20
Agri-silviculture	0.076	272.8	356	25.3	4653	276.8	7.49	4.52
Silvi-pasture	0.067	231.4	296	1.5	4300	245.6	6.06	3.20
Agrislivihorti	0.065	256.7	287	21.5	4404	259.8	5.40	2.90
Eucalyptus	0.051	191.1	221	14.4	2746	242	4.94	2.03
Leucaena	0.062	248.2	256	23.3	3644	245.2	5.18	2.72
Dendrocalamus	0.058	178.3	216	11.7	2653	218.2	3.80	1.98
Aacia albida	0.054	189.2	245	20.8	2781	247	3.80	1.22
Azadirachta	0.044	165.2	218	13.8	2097	179.2	4.30	1.80
Tectona grandis	0.047	178.1	232	10.1	2579	204.5	5.98	1.74
CD (P=0.05)	0.008	12.3	23	3.7	174	14.5	0.52	0.22

Source: Sreemannarayana *et al.*, 2003

Nutrient Cycling under Agroforestry Systems

In a soil-plant system, plant nutrients are in a state of continuous, dynamic transfer. Plants take up nutrients from the soil and use them for metabolic activities. In turn, these nutrients are returned back to the soil either naturally as litter falls in unmanaged systems, deliberately as pruning in agroforestry systems, or through root senescence in both managed and unmanaged systems. These plant parts are decomposed by microbial activities, thereby releasing nutrients into the soil which become available for plant uptake once again (Nair *et al.*, 1999). Natural forest ecosystems of the tropics represent self-sustaining, efficient and "closed" nutrient cycling systems with relatively little loss or gain of the actively cycling nutrients, and with high rates of nutrient turnover within the system. In contrast, most of the agricultural systems represent "open" or "leaky" system with comparatively high nutrient losses. Nutrient cycling in agroforestry systems falls between these extremes (Nair *et al.*, 1995). On unvegetated sites, "pioneer" plants (plants which grow and thrive in harsh, low-fertility conditions) begin the cycling of nutrients by mining and accumulating available nutrients. As more nutrients enter the biological system and vegetative cover is established, conditions for other non-pioneering species become favorable. Pioneers like nitrogen fixing trees tend to benefit other forms of life by boosting fertility and moderating harsh conditions.

In agroforestry systems, a knowledge of cycling of nutrients is essential to understand the way in which the fertility, which is lost due to removal of fuel, fodder, fruits, and grains from the system, is restored by biological process or the deficiency is fulfilled by external inputs. Little information on nutrient cycling in agroforestry systems is available. The living biomass of the system acts as a major storage for essential nutrients and releases them slowly to increase the nutrients efficiency of the system. Three different types of agroforestry systems in the western Himalaya, as worked out by Toky *et al.* (1989), are transitory between forests and agricultural systems. Major pathway for recycling of nutrients in these agroforestry systems is through debris of crops and weeds, tree leaf litter, and external inputs through manure and inorganic fertilizers. Total amount of nutrient added annually (kg/ha^{-1}) to the soil compartment in agrisilvicultural (AS), agri-horticultural (AH), and agrihortisilvicultural (AHS) systems at Solan in the western Himalaya have been shown in Table 7.

Table 7 : Total amount of nutrient added annually (kg ha^{-1}) to the soil compartment in agrisilvicultural (AS), agrihorticultural (AH), and agrihortisilvicultural (AHS) systems at Solan in the western Himalaya

Component	Nutrient (kg ha^{-1})				
	N	P	K	Ca	Mg
AS					
Litterfall	15	1	8	5	8
Crop debris	6	0.2	12	2	5
Weed debris	75	8	69	34	35
Fertilizers	135	80	40	28	19
	(231)	(89)	(129)	(69)	(67)
AH					
Litterfall	31	2	26	15	10
Crop debris	53	6	96	42	53
Weed debris	48	5	43	23	15
Fertilizers	115	80	40	28	19
	(247)	(93)	(205)	(108)	(97)
AHS					
Litterfall	26	2	20	11	10
Crop debris	95	6	103	59	37
Weed debris	48	6	42	27	18
Fertilizers	115	80	40	28	19
	(284)	(94)	(205)	(125)	(84)

Annual turnover of different elements for the soil and vegetation is given in Table 6. Annual turnover of all the elements in the vegetation followed a consistent pattern. Percentage turnover of nitrogen, calcium, and magnesium in the soil compartment was generally much lower than in the vegetation. The turnover rate in the later compartment was much higher, up to 95 per cent for phosphorus and more than 50 per cent for the remaining elements (Table 8).

Table 8 : Fractional annual turnover of elements (the weight that left the compartment divided by the weight held in that compartment, expressed as percentage) of soil and vegetation compartments of agrisilvicultural, agrihorticultural and agrihortisilvicultural systems at Solan in the western Himalaya.

Element (%)	Agrisilvicultural system		Agrihorticultural system		Agrihortisilvicultural system	
	Soil	Vegetation	Soil	Vegetation	Soil	Vegetation
N	3.8	64.9	4.7	73.5	5.2	69.5
P	24.4	89.0	50.5	90.3	52.0	95.4
K	45.2	63.2	65.4	78.9	77.3	73.5
Ca	1.7	51.8	2.5	57.0	3.1	55.8
Mg	5.2	77.0	5.4	66.5	6.0	57.0

The studies on nutrient cycling showed that the introduction of shisham and subabul with crops, significantly increased the organic carbon content over treatments of pure crop and trees alone. The organic carbon status in subabul + maize treatment was 0.72%, followed sisham + maize (0.61%), while in crop alone it was 0.48% only (AICRPAF, 1999-2000).

Nutrients are held in five compartments within an ecosystem *viz.* plant biomass, animal biomass, detritus, soil and rock minerals and available soil nutrients. The movement of nutrients from available soil nutrient pool is induced by root uptake and subsequently, nutrients move to animal and detritus components. In a majority of trees, the nutrient concentration in above ground parts follows the order: leaf>twig>branch>bole; while, in below ground part, the order is fine root> lateral root>stump root. The processes leading to nutrient return to soil from vegetation are falling of plant roots, leaching by rainwater and throughfall. Mehra *et al.* (1985) compared the nutrient return from different forest types and it was found that oak forest added maximum N (76.3 kg), K (49.7 kg) and Ca (102.0 kg); while the addition was minimum in case of pine forest (Table 9).

Table 9 : Nutrient return from different forest types in India

Forest type	Process	Nutrient return (kg $ha^{-1}yr^{-1}$)				
		N	P	K	Ca	Mg
Sal forest	Litter fall	69.8	5.4	31.4	80.7	NA
	Through fall + stem flow	3.8	0.7	6.4	6.0	4.4
	Total	**73.6**	**6.1**	**37.8**	**86.7**	-
Pine mixed broad leaf forest	Litter fall	47.7	4.1	22.8	50.8	NA
	Through fall + stem flow	4.3	0.6	10.0	5.8	5.1
	Total	**52.0**	**4.7**	**32.8**	**56.6**	-
Pine forest	Litter fall	56.9	6.4	37.1	74.2	NA
	Through fall + stem flow	5.2	0.7	12.9	10.5	6.2
	Total	**62.1**	**7.1**	**50.0**	**84.7**	**6.2**
Tilonj dominated mixed oak forest	Litter fall	70.0	5.8	31.9	91.6	NA
	Through fall + stem flow	6.3	0.6	17.8	10.4	6.4
	Total	**76.3**	**6.4**	**49.7**	**102.0**	-

Source: Mehra *et al.* (1985)

The leaves of certain tree species are very rich in mineral composition. Various species of leucaena have a content of 2.3 to 6.3, 0.11 to 0.65, 0.9 to 3.0, 0.2 to 2.1, 0.14 to 0.32 and 0.18 to 1.07% of N, P, K, Ca, Mg and S, respectively. The content of other nutrients varies from 19 to 151, 1.5 to 10.1, 71 to 380, 23 to 172, 104 to 682 and 28 to 50 m g g^{-1} of B, Cu, Fe, Na and Zn, respectively (Dalzell *et al.*, 1998).

Study of nutrient dynamics under 14 years old plantation of *Populus deltoids*, a common tree for plantation on field boundary and agroforestry in loam soil of Dehradun showed a significant addition of nutrients. Litter production consisting of leaves and twigs was 3.8 t $ha^{-1}yr^{-1}$ and it added about 29.4, 29.0, 23.8, 21.3 and 2.8 kg ha^{-1} of N, Mg, Na, K and P, respectively (Dadhwal *et al.*, 1994).

Conclusion

The role of trees on soil fertility and productivity improvement is the focal point in ecosystem management research. The agroforestry interventions are beneficial in increasing biomass production, soil amelioration, stabilization of degraded lands of the country, with special reference to Shivalik region. Moreover, agroforestry options have high potential for sequestering carbon and these could lead to meet the biomass requirement, improvement in soil fertility, conservation of soil and water resources, biodiversity, watershed protection and ultimately sustainable development. As degraded lands of Shivaliks are extremely carbon depleted, there is ample scope of higher carbon accumulation in these lands through introduction of suitable tree and grass species along with proper soil management practices. Improvement of soil fertility under agroforestry systems also depends on the use of appropriate herbaceous species. There should be techniques available for efficient recycling of litter and enhancing the *in situ* soil fertility through use of organic and inorganic inputs at proper time, which will help in increasing the biomass production. Soil quality aspects are to be dealt in detail so as to evaluate soil in various agroforestry systems.

The management of degraded lands of Shivaliks through agroforesty innovations not only meets the multiple needs of food, fodder, fuelwood, fibre, fertilizer etc, but, also ensures economic and environmental security. The adoption of favourable agroforsetry interventions with people's participation on vast areas in Shivalik region through good land use planning and management options would be a step towards sustainable natural resource management in Shivaliks.

References

AICRPAF, (1999-2000). Annual Report of All India Co-ordinated research project on agroforestry. Nauni, Solan (Himachal Pradesh), 45p.

Avery, M.E. (1987). Soil fertility and conservation in agroforestry systems. Proc. Short course on Agroforestry. Colorado State University. Fort Collins (USA): 54-72.

Balkrishan and O.P. Toky (1993). Significance of nitrogen fixing woody legume trees in forestry. Indian Forester: 126-132.

Batjes, N.H. (1996). Total carbon and nitrogen in the soils of the wprld. European Journal of Soil science. 47: 151-163.

Bhatt, B.P., R. Singh, L.K.Misra, J.M.S. Tomar, Matvar Singh, D.S.Chauhan, S.K. Dhyani, K.A. Singh, K.R. Dhiman and M Data (2001). Agroforestry research and practices: An overview. *In*: Verma, N.D. and Bhatt, B.P. (eds.), Steps towards modernization of agriculture in NEH region. pp. 365-392. ICAR Publication, Meghalaya, India.

Bhatt, B.P. (2003). Agroforestry for sustainable mountain development in N.E.H. Region. In Rawat, M.S.S. (ed). *Central Himalaya: Environment and Development (Potentials, Actions and Challenges)*, Published by Transmedia publisher, Srinagar Garhwal, Uttaranchal, India. pp. 206-223.

Bhattacharyya, P., R.P Yadav and Y. Agnihotri (2009). Impact of prevalent silviculture systems on soil stability and potential carbon storage in Shivalik region of lower Himalayas. Journal of the Indian Society of Soil science, 57: 71-75.

Binkley, Dan (1981). Nodule biomass and acetylene reduction rates of red alder and Sitka alder on Vancouver Island, B.C. Canadian Journal of Forest Research 11: 281-286.

Brown, S., C.A.S.Hau, W.Knabe, J.Reich, M.C. Trexler and P. Woomer (1993). Tropical forests: their past, present and potential future role in terrestrial carbon budget. Water and Soil Pollution. 70: 71-94.

Burns, R.C. and R.W.F.Hardy (1975). Nitrogen fixation in bacteria and higher plants. New York: Springer-Verlag.

Dadhwal, K.S., A.Raizada P. Narain and D.P. Datta (1994). Agroforestry systems of Doon Valley and plains of western Uttar Pradesh. *In*: Agroforestry: Traditions & Innovations (eds. Pratap Narain, Dadhwal, K.S. and Singh, R.K.), CSWCRTI, Dehradun, pp. 12-18.

Dalzell, S.A., J.L. Stewart, A. Tolera and D.M. Mc Mell (1998). Chemical composition of *Leucaena* and implications for forage quality. In: Leucaena-Adaptation, quality and farming systems (eds. Sheleton, H.M. Gutteridge, R.C, Mullen, B.F. and Bray, R.A.). pp. 227-245.

Diem, H.G. and Y.R. Dommergues (1990). Current and potential uses and management of Casuarinaceae in the tropics and subtropics. In: Schwintzer, C.R.; Tjepkema, J.D., eds. The biology of Frankia and actinorhizal plants. New York: Academic Press, Inc.

Galloway, James N.; William H.Schlesinger, Hiram II Levy, Anthony Michaels and Jerald L Scnoor (1995). Nitrogen fixation: anthropogenic enhancement-environmental response. *Global Biogeochemical Cycles* 9: 235-252.

Grewal, S.S., Kehar Singh., M.L Juneja and Shiv Charan Singh (1996). Relative growth, fuel-wood yield, litter accumulation and conservation potential of seven acacia spp. On a sloping bouldory soil. Indian Journal of Forestry 19: 174-182.

Gupta, R.K. and D.L.N. Rao (1994). Potential of wasteland for sequestering carbon by reforestation. *Current Science.* 66: 378-380.

IPCC (2000). IPCC special report on land use, land use change and agroforestry. *A Special report of the Intergovernmental Panel on Climate Change.* World Meteorological Organization, Geneva, Switzerland.

Kadirvel, G., B.P. Bhatt, and Anubrata Das (2003). Agroforestry-a promising farming system for livestock production in NEH region.In: Bhatt, B.P. Bujarbaruah, K.M., Sharma, Y.P. and Patiram (eds). Approaches for increasing agricultural productivity in hill and mountain ecosystem. pp. 139-147. ICAR Research complex for NEH region, Umiam, Meghalaya.

Kukal, S.S., H.S. Sur, and S.S.Gill (1991). Factors responsible for soil erosion hazard in sub-montane Punjab, Idia. *Soil Use and management.* 7: 178-180.

Mehra, M.S., P.C. Pathak and J.S. Singh (1985). Nutrient movement in litterfall and precipitation components for central Himalayan forests. Annals of Botany. 55: 153-170.

Mishra, V.K. J.N. Raina, and B.K. Nayak (2000). Multipurpose trees enrich degraded watershed an approach for rural rehabilitation. *Journal of soil and Water Conservution,* 44(1 & 2): 60-71.

Nair, P.K.R., R.J.Buresh, D.N. Mugendi and C.R. Latt (1999). Nutrient cycling in tropical agroforestry systems: Myths and science. In: *Agroforestry in sustainable agricultural systems,* 1–31. Boca Raton, FL: CRC Press.

Nair, P.K.R., B.T. Kang and D.B.L.Kass (1995). Nutrient cycling and soil erosion control in agroforestry system. In *Agriculture and environment: Bridging food production in developing countries* (ASA Special Publication No. 60), ch. 7. Madison, Wisc.: American Society of Agronomy.

Nair, P.K.R. (1984). Soil productivity aspects of agroforestry. ICRAF. Nairobi. 85 p.

Palaniappan, S.P., M.Govinda Swami, and K.Annadurai (1994). Soil fertility aspects of tree farming. In: Thampan, P.K.(ed). Trees & tree farming. Peeky Tree crops Development Foundation, Cochin, pp.123-148.

Rai, P. (1999). Silvipastoral systems -Growing greener and and increasing the production of wasteland. *Wasteland News* 14 : 48-50.

Raina, J.N. and Pardeep Kumar (2000). Physico-chemical characterization and fertility status of some forest nursery soils of district Sirmour in Himachal Pradesh. *The Indian Forester,* 126(6): 657-663.

Rao, B.N. (1987). Watershed management and social forestry.In: Social forestry for rural development. (Eds). P.K. Khosla and R.K.Kohli.ISTS, Solan, pp. 52-64.

Samra, J.S., M.K. Vishwanatham and A.R.Sharma (1999). Biomass production of trees and grasses in a silvopasture system on marginal lands of Doon Valley of north-west India 2. Performance of grass species. *Agroforestry Systems*, 46: 197–212.

Samra, J.S. and Shiv Charan Singh (2000). Silvipasture systems for soil, water and nutrient conservation on degraded lands of Shivalik foothills (sub-tropical northern India). *Indian Journal of Soil Conservation*, 28(1):35-42.

Singh, Gurbachan., J.S.P. Yadav and G.B. Singh (2000). Multipurpose tree species management.In for agricultural production in India. (eds.J.S.P.Yadav and G.B. Singh). International conference on Natural resource management for sustainable production in the 21st century, New Delhi.pp. 481-547.

Sreemannarayana, B., L.G. Giri Rao, and B. Joseph (2003). Nutrients returns and withdrawls from trees in Agrisilviculture systems. Paper presented at 6th Agricultural Science Congress held at IISS, Bhopal, 6-8 Feb. 2003.

Swarup, A., M.C. Manna and G.B Singh (2000). Impact of land use and management practices on organic carbon dynamics in soils of India. *In*: Lal, R., Kimble, J.M., and Stewart B.A. (eds.). Advances in Soil science: Global climate change and tropical ecosystems.CRC Press. Boca Raton. pp. 261-281.

Thind, H. S. (2000). Integration of nitrogen fixing trees in different farming systems for sustaining productivity in Shivaliks. In: Fifty years of research on sustainable resource managementin Shivaliks (eds.) S.P.Mittal., R.K. Aggarwal and J.S. Samra. CSWCRTI Research Centre, Chandigarh. pp. 393-398.

Thind, H.S. and M.S. Dhillon (1994a). Degraded lands of Punjab and their development through agroforestry.In: agroforestry systems for degraded lands (Eds.) Singh, Punjab, Pathak, P.S. and M.M. Roy. Oxford and IBH Publishing Co. Pvt. Ltd. New Delhi, 1:1-12.

Thind, H.S. and M.S.Dhillon, (1994b). Comparative effect of leucaena and eucalyptus on soil fertility and productivity of light textured wastelands. Paper presented at National Conference of Soil and Water Conservation for Sustainable Production and Panchayat Raj held at Chandigarh from January 28-30, 1994.

Toky, O.P., P. Kumar, and P.K. Khosla (1989). Structure and function of traditional agroforestry systems in the western Himalaya. II. Nutrient cycling *Agroforestry Systems*, 9: 71-89.

Tomar, V.P.S., P. Narain and Dadhwal, K.S. (1992). Effect of perennial mulches on moisture conservation and soil-building properties through agroforestry. *Agroforestry Systems*, 19: 241-252.

Vishwanatham, M.K., J.S. Samra and A.R. Sharma, (1999). Biomass production of trees and grasses in a silvopasture system on marginal lands of Doon Valley of north-west India 1. Performance of tree species. *Agroforestry Systems,* 46: 181-196, 1999.

Yadav, R.P., R. Prasad, R.K.Aggarwal, Y. Agnihotri and J.S. Samra (2005). Aonla (*Emblica officinalis*) based silvi-pastoral systems for soil and water conservation in degraded lands of Shivalik foothills. *Indian Journal of soil Conservation,* 33: 207-212.

Chapter-9

Agroforestry for Rehabilitation of Mine spoils in Dehradun Valley

K.S. DADHWAL AND G.P. JUYAL
Central Soil & Water Conservation Research & Training Institute
218 Kaulagarh Road, Dehradun-248 195

ABSTRACT

Mining is an important activity for economic development. However, it is associated with land degradation, pollution of air and water, droughts, floods, land slides and land slips, silting of reservoirs, chocking of bridges and disruption of communications are some of the very serious environmental hazards. The need of the hour is to scientifically extract the mineral coupled with adopting mechanical and vegetative measures to rehabilitate the mine spoils. In the present paper effect of mining on natural resources has been detailed, also techniques adopted and standardized for rehabilitation Sahastradhara mined area of Dehra Dun valley over two decades have been clubbed.

Introduction

Minerals are important raw materials for the development of any nation. But, it is the indiscriminate and unscientific mining activity carried out in the country, particularly in the eco-fragile zones, which has caused irreparable damage to the ecosystem and environment. Himalayan region which constitute about 54 million ha (16.5%) area of the country is characterized by marginality, inaccessibility and fragility, compounded by high intensity erratic rainfall, steep slopes, large scale deforestation and faulty management practices. Natural resources in the region are under great stress due to mounting human and livestock populations. The pristine beauty of Mussoorie hills in Dehradun valley had been badly marred by ugly scars due to large scale indiscriminate mining of high quality limestone for over three decades. The increasing frequency of natural disasters, e.g. landslides, floods, droughts and deterioration of water bodies due to accelerated deforestation, conversion of marginal/forest land into agriculture and unscientific developmental activities have further aggravated the problem.

Though, extraction of minerals is very important for the economic development of any nation, but, their unscientific mining poses a serious threat to the environment (Dadhwal and Katiyar, 1995). Due to excessive use of explosives in mining operations, shear resistance of soil mass is reduced and shear stresses are increased. Due to unscientific mining in hilly terrains, slope height increases, resulting in slope instability. During heavy rains, mass movements are caused by mechanical activation of overland flow and sub-surface water (Haruyama, 1980). Therefore, due to mining, land degradation, pollution of air and water, droughts, floods, land slides and land slips, silting of reservoirs, chocking of bridges and disruption of communications are some of the very serious environmental hazards. In Dehradun valley, there were about 62 mines (active and abandoned) which created land degradation and environmental problems; but in 1985, the Hon'able Supreme Court of India banned quarrying in most of these mines keeping in view the environmental problems. Mine spoil/debris from these degraded sites not only disrupted roads, choked bridges, reduced the capacity of streams, silted reservoirs, polluted air and water, vanished the biodiversity but also affected good agricultural lands nearby. Therefore, the problem of mine spoils required dedicated and concerted efforts for rehabilitation for which research efforts were attempted from 1984 onwards and technologies were developed for their rehabilitation. Techniques developed, applied and their impact are briefed in the following sections.

Limestone Mining in Dehradun Valley (Brief History)

Initiation of collection of limestone in Dehradun valley may be traced back to the second half of the nineteenth century when poor local people/labourers used to search it for the supply to the local lime kilns from the riverbeds of Bindal, Rispana, Tons and Asan, which are the tributaries to holy Ganga and Yamuna rivers. The people took up this job to earn their livelihood on a large scale and this practice subsequently lead for the search of main source of lime in upper hill catchments. One of the contractors, by obtaining license started large-scale extraction of limestone in Dehradun district in 1947 by paying rupees 20 per ton as royalty. After the partition of country in 1947, some of the migrants from West Pakistan settled in Dehradun and applied for permits/licenses for limestone extraction to earn their livelihood.

During the decade 1947-57, *Zamidari* (land lordship) system prevailed. The contractors after fixation of adhoc rate of royalty started the quarrying of limestone, which was continued till 1957. However, the commercial mining of limestone on large scale started from 1960's due to its high off-take by industries like steel, sugar, textiles, medicines, cosmetics etc. This limestone was preferred because of its high grade and purity (92-95%). In 1960 in "Indian Mines and Mineral Act" for quarrying and mining, the rules and regulations were formed in which the State Governments started giving mine areas on lease for a period ranging from 15 to 20 years. Subsequently in 1963, the State Government of Uttar Pradesh framed rules for extraction and quarrying of stones and minerals as "Upkhanij Pariher Niyamawali". Here the rate of royalty was reduced to 30 paise per ton due to the efforts put in by one of the M.L.As and the excess of paid royalty was returned to the respective industrialist/contractor. In a process, the limestone extraction rose to a level of about 1.1 million tonnes during eighties from 62 odd open cast mines in Dehradun valley spread over a length of 40 kilometres and covering an area of about 1400 hectares. All these mining areas were by and large situated on steep and rugged unstable hill slopes between an elevations ranging from 600 m to 2000 m above msl with an average annual rainfall of 2000 mm which is received mostly from July to September.

Factors Conductive for Mine Spoil Wastelands

- Unscientific blasting of more and more areas for mineral exploration.
- Lack of stringent rules and regulations governing environmental controls and precautions.
- Improper planning e.g. not installing diversions, vegetative waterways, etc. well ahead of the actual operation of mining.

Damaging Effects of Mining in Dehradun Valley

The unscientific mining on steep slopes has done immense damage to the fragile ecosystem of the area in the form of heavy soil erosion, drying of water sources, land degradation, disruption to communication systems, floods and consequent decline in food and fodder production in the valley. Limestone mining carried out in the upstream reaches of Doon valley reduced food production by 28%, water resources by 50% and livestock production by 35% (Anonymous, 1988). Some of the environmental damages are listed below:

Soil Erosion and Land Degradation Problems

Devoid of the vegetal cover and land disturbance by mining, the uncontrolled runoff carried the huge amount of soil, boulders and debris from the mined lands since the limestone desposits are highly fractured and fragmentary in nature. The mining operations resulted in huge quantity of undersized over-burden material (about 35%), which have no off-take for industry and thus are dumped as mine rejects or spoils at places. These minespoil materials are highly susceptible to erosion and wash down the hill slopes during the rainy season. The mass soil erosion from the mined areas in Dehradun valley was found to the tune of 550 t/ha/annum (Katiyar *et at.*, 1987) which is much more than the permissible soil erosion limit of 4.5-11.2 t ha^{-1} yr^{-1} and about 3 t $ha^{-1}yr^{-1}$ annum from a good forested area (Dhruvanarayana and Ram Babu, 1983).

The amount of debris coming out of a mere 64 ha Khairawan Dhandola mine at Sahastradhara near Dehradun was measured about 17000 cum/annum. Huge boulders weighing more than 50 tonnes rolled down the hill slopes (Juyal *et al.*, 2007). The debris from the mines of Sahastradhara used to flow to Baldi river downstream which is tributary of river Ganga, joining it near Hardwar. During the last forty years, due to sedimentation, the riverbed has risen by about 15-20 m. Prior to mining operations, it was a perennial river of clean water with a width of hardly 10m. Earlier there were paddy terraces irrigated by this river. Now the river has turned to a torrent, destroying the productive lands along its banks.

Effect on Water Resources

Headwater reaches are the storehouse of rainwater which recharge the aquifers for sustainable water yield to the downstream regions. The limestone belt plays an important role in the hydrology of the region. It temporarily absorbs and stores much of the precipitation and runoff received on the slopes and release it slowly as sub-surface flow, which is the main source of water in Dehradun valley for the domestic and irrigation water requirements. The destruction of this limestone belt by mining activity has adversely affected the water resources in the region. In one of the Doon canals, Mathur *et al.* (1984) observed that the pH, calcium and magnesium contents in the water increased drastically during the peak mining period of February to April thus affecting its quality (Table 1).

Table 1 : Effect of mining on water quality

Particulars	Average content during	
	Peak mining period (Feb-April)	Rest of the period
pH (1 :2.5 soil : water susp.)	8.1	7.3
Calcium (ppm)	150	48.6
Magnesium (ppm)	55.5	3.4

Moreover, due to intensive quarrying around Benog peak, the discharge of Kempty Fall, an attractive tourist spot near Mussoorie, had reduced (Negi, 1982). Similarly, the water falls on the north of Sahastradhara have dried up due to quarrying around the area (Rajwar, 1988). At Hathipaon, Durmala and Kempty Fall areas, there has been a continuous fall in the water table (Katiyar *et al.*, 1990). The substantial damage to the underground hydrological system is also reflected in the fact that spring sources of 18 villages in the Baldi valley have registered an average drop of about 50% in their lean period discharges in the last two decades. Rajpur and Bijapur canal systems which tap water from Rispana and Tons rivers, respectively in the Doon valley, registered a sharp decline of 30% and 40%, respectively in their lean period flows since 1975 (Bandhyopadhyay and Shiva, 1985). The ill effects of quarrying are also reflected in the heavy silting up of rivers and agricultural fields downstream.

In Doon valley, due to mining, many of the perennial streams and rivers have converted to muddy and flash flow torrents causing floods and damage to life and property every year. The torrents emanating from mined headwater regions are damaging about 100 ha of forest land in the valley every year and destroying the trees estimated worth Rs.1 crore (Juyal *et al.,* 1995).

Mining impacts on vegetational biodiversity

Extensive mining for limestone carried out in Doon valley from Rishikesh to Mussoorie has done immense damage to the vegetation resource of the area. Many of the prominent species have become extinct (Malhotra, 1968). Due to washing of top soil and degradation in soil properties (texture, pH, organic matter, N, P, K, water holding and soil depth), the natural bio-diversity has been invaded by homogeneous species of inferior types, such as, *Lantana camara* and *Eupatoreum.* In the area surveyed, it was revealed that as many as 81 species of different trees, shrubs, herbs and grasses have either become extinct or were endangered (Table 2). The high pH due to high calcium content in the soils at Sahastradhara by mining resulted in physiological disorders and yellowing in Eucalyptus plantation, because of hampering of nutrient uptake by plants (Bakshi, 1976). Similarly air pollution by limestone kilns caused a disease of *black tip* in mango plants affecting the fruit production (Raychaudhari *et al.,* 1970). Disappearance of water sources in the mined area of Doon valley have resulted in extinction of at least 20 endemic pterydophytic ferns growing in moist soil conditions (Malhotra, 1968).

With biotic interference and accelerated soil erosion four degraded stages have been recognised by Champion and Seth (1968).

$5B/C_2$-DS_1: It is a low broken soil cover of shruby growth, 3-6 m high, malformed due to biotic interference. The grass cover occurs throughout. Dry deciduous shrub species are: *Woodfordia fruiticosa, Carissa opaca, Mallotus philipinensis, Acacia catechu, Mimosa himalayana, Euphorbia royaleana, Lannea coromandelica, Toona ciliata, Bauhinia retusa, Bauhinia vahlii, Erythrina suberosa, Bombax ceiba* and *Rumex nepalensis.*

$5B/C_2$-DS_2 - Dry savannah: Trees stand far apart singly or in groups or patches with more or less heavy grass. *Phoenix acaulis, Bombax ceiba* and *Toona ciliata* are characteristic species of this type.

$5B/C_2$-DS_4 - Grassland: Aristida cynatha, Saccharum spontaneum, Heteropogon contortus, Cymbopogon nardus and *Eriophorum cosmosum.*

Under rocky phase conditions *Saccharum spontaneum, Carissa opaca* and *Woodfordia fruiticosa* reflect degraded condition of DS3 and DS4 such that the job of revegetation of minespoils with multi purpose trees (MPTs), shrubs and grasses is very difficult and requires suitable techniques and dedicated efforts.

Table 2 : Impact of mining on species richness in Dehradun valley

Species habitat (1)	Species before mining (2)	Extinct (3)	Endangered (4)	Vulnerable (5)	Rare (6)	Threatened (7)
Trees	*Acacia catechu*	-	-	-	-	√
	Bauhinia purpurea	-	√	√	-	-
	B.retusa	-	-	-	-	-
	B. variegata	-	√	-	-	-
	Boehmeria rugulosa	-	√	-	-	-
	Bombax ceiba	-	√	-	-	-
	Brassiopsis aculeata	√	-	-	-	-
	Cassia fistula	-	√	-	-	-
	Celtis australis	-	√	-	-	-
	Dalbergia serica	√	-	-	-	-
	D.sissoo	-	-	-	-	-
	Emblica officinalis	√	-	√	-	-
	Erythrina suberosa	-	√	-	-	-
	Ficus roxburghii	-	-	-	-	-
	F.hispida	-	-	√	-	-
	F.palmata	-	-	√	-	-
	Grewia elastica	-	-	√	-	-
	G. optiva	-	-	√	-	-
	Lannea cormandelica	-	√	-	-	-
	Mallotus philippinensis	-	√	-	-	-
	Moringa oleifera	-	-	√	-	-
	Musa sapientum	√	-	-	-	-
	Olea glandulifera	√	√	-	-	-
	Ougenia oogeinesis	-	√	-	-	-
	Phoenix acaulis	-	-	-	-	-
	Piltosporum eroscarpum	√	-	-	-	-
	Sagertia oppositifolia	√	-	-	-	-
	Sapium insignii	√	-	-	-	-
	Spermadictyon suaveolens	-	-	-	-	-
	Sterculia pallens	-	-	-	-	-
	Toona ciliata	-	-	√	-	-
	Wenlandia exserta	-	-	√	-	-

Shrubs	*Adhatoda zeylanica*	-	√	√	-	-
	Artimisia vulgaris	-	-	√	-	-
	Barlaria cristata	√	-	-	-	-
	Berberis asiatica	√	-	-	-	-
	Carissa opaca	-	-	-	-	-
	Cocculus laurifolius	-	√	√	-	-
	Euphorbia royaleana	-	-	√	-	-
	Lantana camara var. aculeata	-	-	-	-	√
	Leptodermis lanceolatus	√	-	-	-	-
	Mimosa himalayana	-	-	√	-	-
	Syn. M rubicaulis	-	-	-	-	-
	Murraya paniculata	-	-	√	-	-
	Periploca calophylia	√	-	-	-	-
	Randia tetrasperma	-	√	-	-	-
	Rubus ellipticus	√	-	-	-	-
	Rumex nepalensis	-	√	-	-	-
	Sagertia oppositifolia	√	-	-	-	-
	Sophora mollis	√	-	-	-	-
	Tylophora himalaica	√	-	-	-	-
	Vitex negundo	-	-	√	-	-
	Woodfordia fruticosa	-	-	-	-	√
Herbs	*Ageratum conyzoides*	√	-	-	-	-
	Amphicome emodi	-	√	-	-	-
	Blumea aromatica	-	√	-	-	-
	Cassia occidantalis	-	-	-	-	-
	Cassia torga	-	-	√	-	-
	Eupatorium odoratum	-	-	-	√	-
	Euphorbia hirta	-	√	-	-	-
	Fumaria parvjlora	-	-	-	-	√
	Galium rotundifolium	-	-	√	-	-
	Tridex procumbens	-	√	-	-	-
Climbers	*Acacia pennata*	-	√	-	-	-
	Atylosia scarbaeolides	-	-	√	-	-

	Bauhinia vahlu	-	√	-	-	-
	Phaseolus atropurpureus	√	√	-	-	-
	Pueraria tuberosa	-	-	-	-	-
Grasses	*Apluda mutica*	-	-	√	-	-
	Arthraxon tan ceo latus	-	√	-	-	-
	Aristida cyanautha	√	-	√	-	-
	Bothriochloa intermedia	-	-	-	-	-
	Chryspogon fulvus	-	-	-	-	√
	Chrysopogon nardus Var. Strachyii	-	-	√	√	--
	Eragrostella nardoides	-	-	√	-	-
	Eragrostis unioloides	-	-	√	-	-
Sedge Grasses	*Eriophorum cosmosum*	-	-	-	√	-
	Heteropogon contortus	-	√	-	-	-
	Poganotherium paniceum	-	√	-	-	-
	Saccharum spontaneum	-	-	√	-	-
	Themeda anathera	-	√	-	-	-
	Thysanolaena maxima	-	-	√	-	-

Impact on climate change

The Doon valley had been famous for its salubrious climate and natural beauty. Early residents of Dehradun recall with nostalgia those good old days when the outskirts of city used to be lashed by snow during winter months and the entire valley used to receive incessant torrential rains during the monsoon season. Excessive mining, urbanization and industrial growth have deteriorated the environmental health of the valley to a pathetic condition. Not only has snowfall become a thing of the past but also rains have become scantier and erratic. With a large part of forest cover removed, its moderating effect on microclimate is no more. In Doon valley, formerly the summer period having high maximum temperature used to be for a fortnight only, which is now prolonged to more than 1 - 2 months of May to end of June. The lime dust particles near the kilns are a constant source of air pollution, measured at 350 micro grams per cum against the safe level of 100.

Impact on communication

Road communication is frequently hampered by heavy debris flow and landslides from the mined areas. At Sahastradhara near Dehradun, the Public Works Department was spending an amount of more than Rs. 1 lakh every year to clear the debris coming from a 64 ha limestone quarry (Katiyar *et al.,* 1987). The road bridge at Sahastradhara was once washed away due to choking of the river by debris from the mines necessitating for construction of a new one. Disruption to communication also affects the economy of local population, as many of them are dependent on eco-tourism.

Impact of mining on soil charactersitics

The coarse fractured dolomite with coarse sand pebbles formed the fragmental soil in the middle region while fine loamy and sandy alluvial deposits formed deluvial soil which is more favourable and suited to plant growth. In the gully drains/'nalah' basins the landslide/slip debris of weathered shales formed the skeltal soil. To define limitations for the establishment of vegetation, characterization of mine spoils is very important. There are number of factors which limit establishment of vegetation on a mine spoil *viz;* poor physical condition, elemental toxicity, low water holding capacity, alkalinity, high bulk density and poor nutritional status (Dhruvanarayana *et al.,* 1987). The minespoils from Dehradun valley are highly variable from loose to hard and compact at different places. Fresh deposits are loose rocks, schists and of limestone mixed with soil (Dadhwal *et al.,* 1991, 1992). The physico-chemical characteristics of limestone minespoil (< 2mm) collected from 20 mine sites under Mussoorie hills (mean of 52 samples) is presented in Table 3 (Dadhwal *et al.,* 1992). The results reveal that minespoil material is neutral to alkaline in reaction, very poor in nutritional status, sandy loam in texture, high in $CaC0_3$ content and bulk density and low in water holding capacity. The minespoil was also high in infiltration rates as compared to adjacent normal soils because of the texture and other physical characteristics of the debris associated with high lime content (Dadhwal *et al.,* 1989).

Some Other Ill Effects due to Mining

- Compaction of minespoil adversely affects plant establishment and growth by increasing resistance to root penetration and to the emergence of seedlings.

Table 3 : Characteristics of limestone minespoil from Dehradun valley and adjoining normal soils

Characteristic	Dehradun valley (av.of 52 locations)	Sahastradhara (Dhandaula-Khairwail watershed) (av. of 6 locations)	Adjoining normal soils
pH (1:2.5 soil: water suspension)	8.0	8.1	6.9
EC (m mhos cm^{-3})	0.25	0.053	0.06
Organic carbon (%)	0.23	0.25	1.12
Calcium carbonate (%)	63.50	68.10	6.2
Total nitrogen (%)	0.02	0.016	0.14
Available P_20_5 (Kg ha^{-1})	4.04	4.0	20.5
Available K_20 (kg ha^{-1})	56.37	44.12	165
Exchangeable H (me%)	0.012	0.009	-
Exchangeable Al (me%)	-	Traces	-
C:N ratio	11.5	14.1	8.0
Bulk density (Mg m^{-3}) oven dry	1.69	1.75	1.34
Water holding capacity (%)	16.6	-	-
Textural class	Sandy loam	Sandy loam	loam
Sand (%)	67.3	66.6	50.0
Silt (%)	15.3	19.5	24.0
Clay (%)	17.4	13.9	26.0

- Due to lack of normal soil and available soil water, plant growth is restricted.
- Reduced water holding capacity.
- Water pollution and environmental hazards.
- Over burden materials are low in organic matter and other nutrient contents, thus hampers plant growth and their establishment.

- Due to low organic matter, N is the primary limiting nutrient for plant growth. However, a complete fertilizer (N, P and K) is needed for plant establishment.
- Reduced microbial growth and activity.
- Hampering of respiration and enzyme activities.

Rehabilitation Measures

Rehabilitation technology for degraded mine sites consist of several steps, each of them playing an important role. None of the measures in the chain should be missed. Some of the important steps emanated from research studies and experience of two decades are briefed as under:

Mechanical measures

Mechanical measures are the first line of defense and are very important to stop further damage. In highly degraded areas, where establishment of vegetation is very difficult due to moisture stress, movement of debris and rocky material on steep slopes, and absence of fertile soil, mechanical measures are must for controlling erosion, conserving soil particles and helping the vegetation to grow. Studies reveal that engineering structures e.g. cross barriers/check dams, log wood barriers, toe walls, gabions/loose check dams etc. arrest the movement of large amounts of loose rock and debris down the unstable and uneven steep slopes of the area and also through the drainage channels. Some of the important engineering measures, for stabilization of abandoned mined/wastelands emanated are as under:

- Planned placement of overburden in mountainous terrain is very important because it maximises the physical stability of the spoil. The elimination of outer fill slopes and control of outslopes is important for easier establishment of vegetation, spoil stability and for creation of configuration which is more compatible with the natural topography.
- During site preparation, position of water courses should be properly designed and assessed. Water control facilities e.g. channels, diversions and culverts, etc. must be properly designed and maintained. Construction of diversion drains is very much essential to avoid runoff water going to weak zones such as slide face and loose minespoil/debris dumps. Plots should be made depending upon topography.

- There must be a provision of road drains and culverts on mine haul roads. At the same time formation of paths and roads is also essential to transport planting material, manures, compost, water, fertilizer, etc.
- There should be construction of continuous contour trenches on minespoils and staggered contour trenches on denuded long slopes. These trenches are useful in water and finer particles retention and establishment of vegetation.
- Terracing of minespoil/debris deposited in heaps of steep slopes is advocated. Terracing will ease further to adopt biological means for minespoil stabilisation.
- Measures like use of Geo-jute, Netlon, Geo-girds, Excelsior matting, mulching, etc. are also important to avoid rain drop impact on loose debris and also to facilitate the revegetation of the area. In steep slopes, the importance of these measures further increases. Mulching helps in moisture conservation and after decomposition building nutritional status.
- Upto 40% slope, the construction of gabion drop structures in the main drainage channel is recommended.
- There must be a construction of silt detention basins in the main drainage channel wherever, technically feasible, normally upto 20% slope.
- In small tributaries (first order streams) construction of logwood, brush wood and loose rock filled check dams is essential.
- Construction of stone filled log wood crib structures on steeply sloping, loose areas especially on slide faces is recommended. Studies reveal that these structures suit better than gabion drop structures where slope exceeds 40%.
- Construction of chute spillways with cement concrete where the vertical fall is more than 3m is advocated.

The above mentioned strategies/measures are useful in grade stabilization, formation of bench terraces, retention of good soil, reduction in the steepness of the channel slopes, retention of runoff and thus better soil moisture status in the soil profile, increased subsurface flow and facilitation in the establishment of vegetation.

Agroforestry interventions

Agroforestry measures must be undertaken alone/simultaneously with the construction of engineering structures depending upon the site and situation. Measures adopted should have positive effect on soil properties and plant growth. The affected areas should be put to productive use, may be agriculture, forestry/agroforestry, grassland or recreation. Aesthetic value should be restored and adverse visual impact be avoided. While taking agroforestry measures, economic and ecological considerations must be kept in mind. The following are some of the techniques emanated after research studies for the rehabilitation of abandoned limestone mined waste lands from Dehradun valley :

1. Since minespoils of Dehradun valley are very poor in nutritional status (pH 8.0, organic carbon 0.13%; available phosphorus 4 kg ha^{-1}; available potash 44 kg ha^{-1}; total nitrogen 0.02%), minespoil/ debris fertility management should be given due importance for the quick establishment of vegetation. Studies conducted for the minespoil, from Sahastradhara (near Dehradun) reveal that 50 kg N $ha^{-1}$and 20 tonnes FYM/Leaf litter ha^{-1} to *Eulaliopsis binata* (bhabhar) gave better results. Similarly, di-ammonium phosphate @ 10 g/tree sapling gave encouraging results in the first year of establishment of MPTS.
2. Leaf litter/mulches are useful in moisture retention and fertility buildup. Leaf litter collected from the nearby forest (composed of grasses, wild Kadzu, *Sapium insignii* leaves etc.) after chopping incorporated in minespoil/debris gave better results for bhabhar growth and 20 t ha^{-1} was found to be the ideal dose.
3. Vegetation establishment can be enhanced by placing and mixing of a normal soil with minespoil. One can have a normal soil from adjacent good areas. Placing normal soil upto a depth of 30 cm in the initial year of establishment was found good for *Eulaliopsis binata* (bhabhar) and *Chrysopogon fulvus* (gorda) grasses whereas in second year mixing treatment upto the same depth was found better over placing. Mixing 50:50 (soil : debris) performed well for the establishment of various tree, shrub and grass species. These studies suggest that if normal soil is available nearby then that can be used in pits/holes/channels and can be mixed with debris before planting for better establishment of vegetation.
4. Select preferably leguminous plants for such degraded sites, which may add nitrogen. For example *Leucaena leucocephala*

(Subabul) and *Pueraria hirsuta* (Kudzu) have root nodules which fixes nitrogen. The establishment of subabul and its growth is quick and can be used as a fuel and fodder (mixed with other grasses). Legumes can also be mixed with other plant species for re-vegetation of such affected sites.

5) While selecting plant species for rehabilitation of abandoned mined lands the following considerations should be taken into account:

   - Plant species selected should provide quick and immediate vegetative cover on these waste/barren and affected sites to avoid further damage. For immediate cover, grasses are best proposition. Grasses like *Chrysopogon fulvus* (gorda), *Eulaliopsis binata* (bhabhar), *Pennisetum purpureum* (hathighas), etc. have given good performance at Sahastradhara near Dehradun. *Bhabhar* is an industrial grass whereas other two can be used as animal fodder.
   - Species selected should have economic and ecological values.
   - Should add nitrogen (preferably).
   - Since abandoned limestone sites are very rich in calcium carbonate content ($CaCO_3$ varies 4-80%), plant species selected should be lime loving.
   - Plant species chosen should have better root soil binding characteristics.
   - Should have aesthetic value.
   - Plant species chosen should be preferably native of that place.

6) Biological measures like contour wattling, mulching and planting can be adopted for such affected sites depending upon slope and other site characteristics.

7) In highly steep lands spraying over the dumps with seeds mixed with cowdung, hay, made into slurry, etc. is suggested.

8) Studies conducted in Dehradun valley reveal that plant species like *Chrysopogon fulvus* (gorda), *Eulaliopsis binata* (bhabhar), *Pennisetum purpureum* (hybrid napier), *Pueraria hirsuta* (kudzu wine), *Saccharum spp* (kans grass), *Agave americana* (rambans), *Ipomoea carnea* (sadabahar/besharam), *Eucalyptus* hybrid (eucalyptus/safeda), *Bauhinia spp.* (kachnar), *Leucaena leucocephala*

(subabul), *Vitex negundo* (semalu), *Salix tetrasperma* (salix/ jalmala), *Acacia catechu* (khair), *Erythrina suberosa* (Jhingan), etc. have performed well and can be used for the rehabilitation of such abandoned ruined sites. The various agroforestry species coming under secondary succession alongwith evaluated and successfully reforested species in different degraded minesites of outer Himalaya are listed in Table 4 (Dadhwal *et al.*, 1997).

Table 4 : Prominent agroforestry species for rehabilisation of limestone minespoil from Dehradun valley with their economic use

Species	Economic Use
TREES	
*Acacia catechu (Khair)**	Wood for K, C, and F; B for tanning
Bauhinia purpurea (Gural)	Fd, wood for small timbers, B for tanning, flower buds as vegetable
B.retusa (Semla)	B, gum for sweets and medicines
B.variegata (Kachnar)	B for medicines; wood as F and small timbers
Boehmeria regulosa (Genthi)	Wood for making bowls and dairy utensils
Bombax ceiba (Semul)	Soft wood species, premature calyx as vegetable, cotton to stiff pillows and cushions, seeds as nourishing cattle feed
Cassia fistula (Amaltas)	Fruit as medicine, B for tanning/dyeing and wood for minor implements
Celtis australis (Khirik)	Fd, fuel and wood for handles and churnsticks
Dalbergia sissoo (Shisham)	F, Fd and timber; SC/afforestation
Dendrocalamus strictus (Bamboo)	Fd, SC, T
Erythrina suberosa (Madara)	Wood for sieve frames, SC
Eucalyptus hybrid *(Safeda)*	F, T, paper pulp
Ficus auriculata (Timla)	Fd, fruits (edible)
F.hispida (Kaksa)	Fd, yound fruits made into carry, B, FR for ropes
F.infectoria (Pilkhai)	Fd, yound fruits made into carry, B, FR for ropes
F.palmata (Pheru)	Fd, fruits (edible)
F.rumphii (Pilkhan)	Fd, fruits (edible)

F.semicordata (Khain)	Fd, fruits (edible), old leaves used in wood polishing
Grewia optiva (Bhimal)	Fd, FR, minor timber, F
Lannea coromandelica (Jhingan)	Fd, gum, SC
Leucaena leucocephala (Subabul)	Fd, F, N-fixation, SC
Mallotus philipinensis (Raini)	F/small timber, B for tannning, medicinal
Moringa oleifera (Sanjan)	Fd, fruit as vegetable, gum, medicine, SC
Murraya exotica (Bilgar)	Wood used in making handles of implements and walking sticks, twigs used as tooth brushes
Ougenia oojeinensis (Sandan)	Fd, implements, furniture, gum
Phoeni acaulix (Khajoori)	Leaves for mat making, broom making, fruits (edible)
Salix tetrasperma (Jalmala)	Wood for gun powder, Charcoal, SC
Toona cilata (Toon)	Fd, flowers fordye, bark medicinal, timber
Wendlandia exserta (Bathua)	Timber, agricultural implements
SHRUBS	
Agave species (Sisal)	Ropes, leaf colloid used in detergent, SC
Adhatoda zeylanica (Vasaka)	Leaves and twigs medicinal
Barleria cristata (Jhinti)	Roots/leaves medicinal, an antidote to snake bite
Carissa opaca (Karonda)	Fd, wood for combs, fencing and burning, fruit edible
Cocculus laurifolius (Tildhera)	Fuel
Euphorbia royaleana (Thor)	Live hedge fence, SC
Impmoea carnea (Besharam)	SC, fuel
Lantana camara (Lantana)	F, furniture
Mimosa himalayana (Al)	Fd, heartwood for tent pegs
Rumex nepalensis (Chulmora)	A weed on calcareous soil
Vitex negundo (Shimalu)	Branches for basket making, roots and leaves medicinal, SC
Wood fordia fruiticosa (Dhaula)	Wood for minor implements, flower for silk dye, and B for medicine

GRASSES AND SEDGES	
Andropogon nardus (Andropogon)	SC
Apluda mutica (Apluda)	Fd
Arundo donax (Narkul)	SC, toys, pipe
Aristida cynantha (Arsitida)	SC, brooms
Bothriochloa intermedia (Bothriochloa)	Fd, SC
Chrysopogon fulvus (Gorda)	Fd, SC
Cynodon dactylon (Doob ghas)	Fd, SC
Eragrostis nardoides (Eragrostis)	SC
Eulaliopsis binata (Bhabar)	Fb, paper pulp, SC
Heteropogon contortus (Speargrass)	Fodder, SC
Pennisetum purpureum (Napier grass)	Fodder, SC
Saccharum spontaneum (Kans)	SC, bufalo Fd, thatching
Thysanolaena maxima (Broom grass)	SC, broom
Eriophorum cosmosum (Pseudobhabar)	Fd, ropes, paper, SC
HERBS	
Argemone mexicana (Pili-katnari)	SC and medicinal
Euphorbia hirta (Dudhli)	Medicinal, weed of dry localities
Eupatorium adoratum (Eupatorium)	Medicinal, a weed in moist habitat
Galium rotundifolium (Galium)	A weed in moist habitat
CLIMBERS	
Acacia pinata (Agai-bel)	Endemic to moist gullies/ravines
Atylosia scarabaeoides (Atylosia)	Fd, SC
Bauhinia vahlii (Malo)	Leaves as plates, cups, table cloth and raincaps; B for ropes, roasted pods and seeds edible
Peuraria hirsuta (Kudzu)	Fd, SC, Organic manure, nitrogen fixing

C=Charcoal, F=Fuel, Fd=Fodder, FR=Fibre, K=Katha, B=Bark, SC=Soil Conservation, T=timber * Common/Vern. names in parenthesis

Mechanical and agroforestry measures applied have given encouraging results for the stabilisation of abandoned limestone mined lands from Doon valley over the time. Impacts of the technologies applied are enumerated as under:

Impact of Technologies on Rehabilitation

Impact on Vegetation

With continued biotic protection together with rehabilitation measures and protection against grazing and cutting of grasses, shrubs and MPTs the retrogression and erosion has come to stand still. The vegetal cover has increased from 10% to > 90% due to reforestation over a period of 25 years.

Impact on minespoil characteristics

Impact of rehabilitation measures on some important characteristics over time was seen in treated area. On an average, after nearly two decades, pH of the spoil came down by 0.7 units (8.1 to 7.4), organic carbon increased from 0.13 to 0.45%, whereas $CaCO_3$ (%) decreased from 54.6 to 29.5 and bulk density (Mgm^{-3}) from 1.63 to 1.45 which shows that planting of MPTs, shrubs and grasses over time have helped in improvement of minespoil characteristics (Table 5).

Table 5 : Changes in minespoil characteristics overtime (average of 14 locations)

Properties	Year						
	1985	1991	1992	1994	1995	1997	2002
pH (1:2.5)	8.1	7.9	7.7	7.6	7.6	7.5	7.40
Organic C (%)	0.13	0.18	0.26	0.39	0.4	0.42	0.45
Total N (%)	0.01	0.02	0	0	0	0.05	0.05
$CaCO_3$ (%)	54.6	34.6	34	32	31.8	31	29.50
Bulk density ($Mg\ m^{-3}$)	1.63	1.53	1.5	1.48	1.48	1.47	1.45

Impact on slope and runoff

The equivalent slope in the treated hilly terrain reduced from 38.3% in 1983 to about 19% in 1997 over a period of 14 years. Similarly, runoff and soil loss also decreased considerably over the time with mechanical and biological measures.

Impact on base flow

Because of the various treatment measures, the main drainage channels which used to dry up by the end of October, have now become perennial. The dry weather flow recorded in the main drainage channel was to the tune of 265 and 100 cum/day in the months of November and February, respectively. Farmers of the area have started utilizing this augmented flow for irrigation purposes.

Impact on water quality

Runoff samples collected from treated and untreated mined areas were analysed for calcium, magnesium, sulphates and bi-carbonates (Table 6). Runoff water from treated minesite was found within the permissible water quality limit (Juyal *et. al.* 1993).

Table 6 : Water quality parameters (ppm)

Site	Ca	Mg	SO_4	HCO_3
Treated oldmine	74	34	138	131
Untreated oldmine	188	39	240	98
Untreated fresh mine	389	120	756	77

Conclusion

It is inferred that as a result of agroforestry and mechanical measures, the soil erosion was checked, the surface runoff was reduced, the lean period flow increased due to ground water recharge, water quality and biodiversity improved (Table 7).

Table 7 : Impact of rehabilitation measures

Particulars	Pre-treatment	Post-treatment
Minespoil outflow (t/ha/yr)	550	6
Surface runoff (%)	57	37
Lean period flow (days)	60	>240
Water quality	Not portable	Portable
Vegetal cover (%)	10	>80
Cost of debris clearance (Rs./yr)	1 lakh	Nil

Future Strategies

Several long-term studies has been conducted for the rehabilitation of mined lands in developed countries, but in India, barring few, very meagre information for rehabilitation of minespoils of various mines is available. Since the rehabilitation measures vary with the type of minespoil, site and physiographic conditions, the following future line of work is suggested for rehabilitation of minespoils from Dehradun valley.

- Quantification, characterization and classification of minespoils.
- To develop site specific soil working and planting techniques.
- To develop site specific fertility management programmes.
- For steep slope stabilization, measures like Geojute, Netlon, Excelsior matting, etc. with planting of suitable vegetation.
- Selection of only such type of plant materials (MPTs, shrubs, grasses, creepers/ climbers, etc.) which provide immediate surface cover, add nitrogen, have good root soil binding characteristics and ecologically and economically sound.

References

Anonymous (1988). *Bulletin No.2/88,* National landuse and Conservation Board, Ministry of Agriculture, New Delhi.

Bakshi, B.K. (1976). *Forest Pathology,* ICFRE, Dehradun, pp.204-205.

Bandhyopadhyaya, J. and V. Shiva, (1985). The conflict over limestone quarrying in Doon valley, Dehradun, India. *Environmental Conservation,* Elsevier Sequoia, S.A., Lausanne, Switzerland, 12(2) : 131-130.

Champion, H.G. and S.K. Seth, (1968). *Forest types of India,* Govt. of India publication, pp. 200-209.

Dadhwal, K.S., P.Narain, B.Singh, and R.Singh, (1989). Infiltration characteristics of limestone minespoil from outer Himalayas of Uttar Pradesh. *J. Indian Soc. Soil Sci.,* 37(2) : 223-228.

Dadhwal, K.S., B.Singh and P. Narain, (1991). Effect of placing and mixing of a normal soil on the performance of two grasses in limestone minespoil. *J. Indian Soc. Soil Sci.,* 39(4) : 751-756.

Dadhwal, K.S., B.Singh, V.S. Katiyar and V.K. Dwivedi, (1991). Characteristics of limestone mined lands. *Indian J. Soil Cons.,* 19(3): 79-82.

Dadhwal, K.S., B. Singh and P.Narain (1992). Characteristics of limestone minespoil/ debris from outer Himalayas of Uttar Pradesh, *Indian Forester,* 118(9) : 650-658.

Dadhwal, K.S. and V.S. Katiyar (1995). Land degradation and environmental pollution by mining and land slides. Bull. No.17, *Indian Soc. Soil Sci.,* pp.114-126.

Dadhwal, K.S., P.Joshie and G.P. Juyal (1997). Prominent agroforestry species for reclamation of limestone minespoils of outer Himalaya. *Indian J. Soil Cons.,* 25(1): 33-40.

Dhruva Narayana, V.V., V.S. Katiyar and K.S. Dadhwal (1987). Mined area reclamation in Mussoorie Hills. *Indian J. Soil Conservation,* 15(2) : 82-90.

Dhruva Narayan, V.V. and Ram Babu (1983). Estimation of soil erosion in India. *J. Irrigation Drainage Engg.,* 109(4) : 419-434.

Haruyama, M. (1980). *Proc. Int. Symp. Landslides,* New Delhi.

Juyal, G.P., V.S.Katiyar, K.S. Dadhwal, and P.Joshie, (1993). Rehabilitation of a steep mined area in Mussoorie hills. *Wasteland News,* 9(1): 57-91.

Juyal, G.P., V.S. Katiyar, and R.K. Arya, (1995). Trapezoidal flume for gauging of hilly watersheds. In : *Torrent Menace – Challenges and Opportunities* (Eds. Sastry *et al.)* CSWCRTI, Dehradun, pp. 164-185.

Juyal, G.P., V.S. Katiyar, K.S. Dadhwal, P.Joshie and R.K. Arya (2007). Mined area rehabilitation in Himalayas – Sahastradhara experience, *Bulletin No. T-53/D-34,* CSWCRTI, Dehradun, 104 p.

Katiyar, V.S., K.S. Dadhwal, S.K. Dhyani, P.Joshie and R.K. Arya, (1987). Sahastradhara mined area rehabilitation project. *Tech. Bulletin,* CSWCRTI, Dehradun, 25 p.

Katiyar, V.S., G.P. Juyal and K.S. Dadhwal (1990). Management of a mined watershed. In : *Third IWRS National Symposium on Watershed Development and Management,* Kanpur, pp. 110-117.

Malhotra, S.K. (1968). Vegetative exploration of the limestone area of Sahastradhara, Dehradun. *Ph.D. Thesis,* University of Rajasthan, Jaipur.

Mathur, H.N., P. Soni, S. Naithani, O. Kumar and H.B. Vashistha (1984). Environmental impact on resource utilization in Doon valley – A few case studies. Paper presented at Symp. on Resource Identification and Utilization in Doon valley organized by *Doon Science Forum.*

Negi, S.S. (1982). Environmental problems in the Himalaya. *Bishen Singh, Mahendra Pal Singh,* Dehradun, India, 110-113 pp.

Rajwar, G.C. (1988). Limestone quarrying – A hazard to Mussoorie mountains and their ecosystem. In : *Mining and environment in India* (Eds. S.C. Joshi and G. Bhattacharya), Himalayan Research Group, Tallital, Nainital, India, pp. 294-298.

Ray Chaudhari, S.P., R. Prasad and D.N. Srivastava (1970). Agricultural Year Book – New Vistas in Crop Yields, ICAR, New Delhi, 449 p.

Chapter-10

Bhimal (*Grewia optiva*) Based Land Use Systems for Resource Conservation and Fodder Severity in Shivalik Himalayas

HARSH MEHTA AND K.S. DADHWAL
Central Soil and Water Conservation Research and Training Institute,
218 Kaulagarh Road, Dehradun 248195

ABSTRACT

Fodder availability is a serious concern during winter season in hilly areas as land for growing fodder crops is hardly available unlike plains. Bhimal as an agroforestry species is being cultivated in the Himalayas since long, however, genetic improvement of the species was hardly undertaken in the past barring a few reports on tree improvement of the species. Seeds were used to be usually collected from crooked trees which are easy to climb or swept from the forest floor. Since agroforestry systems have great promise to improve the productivity of both agricultural and tree crops, therefore, keeping that in view, seven provenances (I.C. Bhaintan , I.C. Chamba, I.C. Katkor, I.C. Malas, I.C. Nauni, I.C. Naur, I.C. Tachla) were collected from different regions of Uttarakhand and Himachal Pradesh and evaluated for 10 years (1995-2004). Significant differences in the productivity potential of Bhimal provenances were noticed and the high yielding fodder provenances identified (I.C. Bhaintan , I.C. Chamba and I.C. Malas), yielding as high as 35 q ha^{-1} thus fodder productivity could be increased to 25% by planting high yielding provenances. Likewise high yielding provenaces (HYPs) for fuelwood and fibre were identified which can increase the productivity potential of the agroforestry system. In 2005, more than 1200 saplings of HYPs were multiplied in a polyhouse and planted at different elevations in the Himalayas under prevailing agroforestry systems. Differential responses of HYPs in respect of growth and productivity at different locations have been monitored. The rooting behaviour of HYPs was also investigated and it was found that all the three provenances viz., I.C. Bhaintan , I.C. Chamba and I.C. Malas have high soil binding characteristics, thus revealing that apart from production, this tree species plays an important role in soil conservation. Further, high regression of soil binding on above ground biomass and specific root length was recorded among these provenaces. Looking at the performance of such provenances on the farmers' fields, the demand of high yielding provenaces is increasing. Unlike Eucalyptus, Bhimal had the minimal effect on the performance of field crops and the overall system productivity of the Grewia optiva based agroforestry system was profitable along with insurance against aberrant monsoon conditions.

Introduction

India's population is increasing at alarming rates. It is expected to cross 1.3 billion by the year 2020. Besides, India has 15 per cent of world's livestock population which depends upon 2.4% of world's geographical area. It exerts tremendous pressure on limited natural resources. The total livestock population (cattle, buffaloes, sheep, goat, equine and camel) in 2010 is estimated at 528.2 million which is expected to grow to 590. 2 million by 2025 (Table 1). The annual requirement of green and dry fodder in 2010 was estimated at 1061.4 and 589.6 million tons, respectively which is expected to rise to 1097.6 and 609.8 million tons, respectively by 2015. It will lead to a net green and dry fodder deficit of 63.5 and 23.6 million tonnes, respectively by 2015 (Table 2).

The scope of increasing area under fodder crops is limited due to pressure on cultivated land for production of food and cash crops to meet the demand of ever increasing population. Efforts have to be made to increase the productivity of fodder crops and tapping alternative sources of fodder. A number of trees found in different agro-climatic regions have enormous potential for producing nutritious green fodder (Rekib *et al.*, 1991). Fodder trees can produce as much green fodder per unit area as agricultural crops (Singh, 1982) and are rich in protein, calcium and vitamins. In hilly and desert areas livestocks are mainly maintained on tree lopping (Deb Roy *et al.*, 1980; Raghvan, 1990). Grasses become dry with the onset of winter and the foliage of fodder trees forms an excellent nutritive fodder during the lean season (Negi, 1977) in the North West Himalayas when other fodders are not available. In a recent study in Central Himalayas, a population density of 334-418 fodder trees ha^{-1} was observed in the agricultural fields which are the main source of fodder in winter (Semwal and Maikhuri, 1996).

Table 1 : Supply and demand of green and dry fodder projected estimates (million tonnes)

	Supply		Demand		Deficit as % demand	
Year	Green	Dry	Green	Dry	Green	Dry
2010	395.26	451.28	1061.4	589.6	62.76	23.46
2015	400.60	466.16	1097.3	609.8	63.50	23.56
2020	405.95	473.79	1134.2	630.1	64.21	24.81
2025	411.30	488.32	1170.80	650.4	64.89	24.92

Source: National Livestock Policy Perspective, 1996

Table 2 : Projected livestock population estimates* (Based on livestock population data from 1950-92)

Year	Cattle	Buffalo	Sheep	Goat	Equine	Camel	Total
2010	226.4	100.6	53.9	145.7	0.3	1.3	528.2
2015	232.2	105.6	55.3	154.3	0.2	1.4	549.2
2020	238.0	110.7	56.7	162.9	0.2	1.4	569.2
2025	243.8	115.7	58.1	171.6	0.2	1.5	590.2

Source: Livestock census, Dept. Animal Husbandry and Dairy, Ministry of Agriculture, New Delh (1996)
**The figures were projected to 2025 using growth trends.*

Grewia optiva locally known as Bhimal, a multipurpose tree species producing fuel, fodder and fibre, is the most popular agroforestry tree in the hill and mountainous agro-ecosystems (Singh, 1982; Joshi and Narain, 1992). It comes up naturally on the terrace risers and the farmers protect and tend the Bhimal seedlings (Dadhwal *et al*, 1989). It can also grow in most degraded sites due to its high soil binding capacity. The species has been identified as an important component of the silvipastoral system for riverbed bouldery lands as well as Himalayan foot hills (Viswanatham *et al.*, 1999).

It is an important small to medium sized tree distributed throughout the Himalayas upto an elevation of 2000 m above msl. It is maintained on terrace risers and other vacant lands around agricultural fields as a component of agroforestry system primarily for leaf fodder. Plants are heavily lopped for their nutritious and palatable fodder, fuel wood and bark for rope making (Photo 1, 2,3, 4, and 5).

Photo 1 : Fodder is a serious problem in hilly regions. Farm women of Uttarakhand bringing a head load of fodder for animals from long distances in the interior Himalayas

Till date no emphasis has been given to the use of high quality seed source for forestry plantations and characterizing the productivity of provenances obtained from different geographical areas. Generally, the seeds are collected from crooked trees which are easy to climb or swept from the forest floor.

Photo 2 : Bhimal is the most favourite fodder tree in the Himalayas, heavily lopped for its nutritious fodder.

Photo 3 : Unscientifically lopped Bhimal trees

Photo 4 : Bark fibres are put to variety of uses. A village family busy in making ropes from bark fibre after retting.

Photo 5 : Debarked branches of Bhimal used as fuelwood.

Improvement Work on Bhimal at Dehradun

Investigations were undertaken with a view to (1) identify suitable provenances of *Grewia optiva* for agroforestry system, (2) assessing the nature and magnitude of genetic variances in a collection of *Grewia optiva* provenances collected from different regions of north-western Himalayas in respect of different growth parameters and productivity of fodder and fuel wood production, and (3) assessing the nature and magnitude of phenotypic and genotypic associations among different growth and productivity parameters to develop suitable selection criteria and define appropriate ideotypes for agroforestry system of the north-western Himalayas. Characteristics of locations, growth parameters and phenological traits of *Grewia optiva* provenances and yield are given in Tables 3, 4 and 5 respectively.

Table 3 : Physical and edaphic features of locations from where provenances have been collected

Provenance	Elevation (m asl)	Soil type/series
I.C. Bhaintan	1225	Moderately gravelly, non calcareous, medium textured, deep to very deep. Well drained with moderate permeability, pH 7.7 in all soil horizons.
I.C. Chamba	1500	Silty clay to silty clay loam in texture. Soil depth upto 2 m with soil pH 5.5 to 5.6
I.C. Malas	1450	Slightly to moderately gravelly, non calcareous, medium textured and very deep soils derived from shale and colluvium of varying depths. Well drained with moderate permeability, pH decreasing (6.7- 6.3) with profile depth
I.C. Katkor	1550	Moderate to very gravelly, stony, non calcareous, medium textured, moderate deep soils derived from slate, fine sand stone, quartzite conglomerate and their colluvium of varying depths, soil occurs on steep slopes at higher altitudes. Well drained soil with moderate permeability with pH 5.0 to 5.4
I.C. Nauni	1250	Sandy loam, moderately deep to deep with coarse medium texture admixtured with gravels and pebbles. Parent material comprising ferro-magnesium shales and dolomitic limestones with soil pH ranging from 6.5 to 7.2.
I.C. Naur	1350	Slight to moderate gravelly. Moderate deep, stony quartzite conglomeration
I.C. Tachla	1300	Rocky with sand stone, quartzite, hard shale and occurs on steep slope.

Table 4 : Initial (1996) and final (2004) growth parameters and phenological traits of different *Grewia optiva* provenances

Provenances	Plant height (m)		Diameter at breast height (cm)		Crown length (m)		Crown width (m)		Branches plant^{-1}		Crown width : bole diamter		1000 fruit weight (g)		Fruit production (g tree^{-1})		Litter fall (ton ha^{-1})	
	I_o	F_o	I_o	F_o	I_o	F_o	I_o	F_o	I_o	F_o	I_o	F_o	I_o	F_o	I_o	F_o	I_{97}	F_o
I.C. Bhaintan	2.62	5.63	1.46	4.74	2.27	4.11	2.22	2.75	32	21	1.532	58.0	-	10.7	-	682	6.83	6.32
I.C. Chamba	2.98	5.48	1.31	4.50	2.28	3.98	2.54	2.78	22	20	1.94	62.0	-	105.6	-	628	5.25	5.85
I.C. Katkar	3.03	5.33	1.49	4.30	2.31	3.81	2.35	2.99	24	21	1.56	70.7	-	100.1	-	622	5.00	5.77
I.C. Malas	2.86	5.86	1.61	4.87	2.65	4.30	2.56	2.86	31	26	1.59	59.0	-	101.7	-	658	6.13	6.82
I.C. Nauni	2.27	5.65	1.40	4.28	2.16	3.85	2.12	2.94	22	18	1.51	68	-	107.8	-	527	4.84	5.48
I.C. Naur	2.72	5.51	1.81	4.55	2.44	4.22	2.11	2.95	23	20	1.17	65	-	109.2	-	570	5.10	5.29
I.C. Tachla	2.59	6.22	1.53	4.99	2.36	4.30	2.46	3.19	33	24	1.61	66	-	106.8	-	5.38	6.00	6.27
X	2.72	5.67	1.51	4.60	2.35	4.08	2.35	2.93	27	21	1.59	64	-	105.0	-	603	5.59	5.97
S.E. (d)	0.1	0.2	0.2	0.1	0.1	0.2	0.1	0.1	1.7	0.9	0.2	1.0		1.4	-	10.2	0.24	0.23
CD (0.05)	0.4	0.6	0.5	0.5	0.4	0.5	0.4	0.2	5.3	2.7	0.6	3.0	-	4.4	-	31.4	0.73	0.71
CV	7.8	6.3	15.8	6.7	8.7	7.3	9.7	4.9	11.2	5.3	23.3		-	2.3	-	2.9	3.4	2.8

Table 5 : Dry weight of lopped fodder (q ha^{-1}), wood biomass (q ha^{-1}) and fibre yield (q ha^{-1}) of different *Grewia optiva* provenances (2001-04)

Provenances	Dry weight of lopped fodder (q ha^{-1})				Dry weight of debarked fuel wood (q ha^{-1})				Dry weight of bark fibre (q ha^{-1})			
	2001	2002	2003	2004	2001	2002	2003	2004	2001	2002	2003	2004
I.C. Bhaintan	13.55	7.96	13.47	15.40	11.78	9.05	6.57	7.14	4.11	4.80	3.76	3.73
I.C. Chamba	9.98	8.76	1322	14.59	12.22	10.23	5.53	5.60	3.68	4.14	3.07	2.83
I.C. Katkor	9.95	7.10	12.80	13.65	10.12	9.18	4.98	5.56	3.28	3.68	2.73	2.39
I.C. Malas	9.50	8.28	13.46	**15.45**	15.73	10.24	5.41	5.78	5.56	3.44	3.03	2.93
I.C. Naur	10.04	7.21	11.04	12.48	10.43	8.59	5.71	4.65	3.56	3.97	2.64	2.34
I.C. Nauni	8.08	5.90	11.55	12.78	7.79	7.65	4.53	4.61	2.59	2.88	2.43	2.02
I.C. Tachla	8.80	7.11	11.94	13.76	9.38	8.63	4.02	3.95	3.16	3.21	2.21	1.86
X	9.98	**7.47**	12.50	**14.02**	11.06	9.08	5.11	5.32	3.70	3.73	2.84	2.59
S.E. (m)±	0.18	0.40	0.21	0.23	0.17	0.66	0.25	0.87	0.03	0.34	0.14	0.40
CD (0.05)	0.55	1.24	0.65	0.71	0.52	2.05	0.77	2.70	0.09	1.05	0.42	1.25
CV	3.09	9.33	2.92	2.83	2.66	12.62	8.44	2.84	2.58	15.80	8.40	2.72

Productivity parameters

This species is primarily grown for its nutritious fodder. Significant differences detected among provenances for leaves and edible twigs ranged between 29.44–35.66 q ha^{-1} during 2001-2004 (Fig. 1). The average values of different productivity parameters after 10 years of

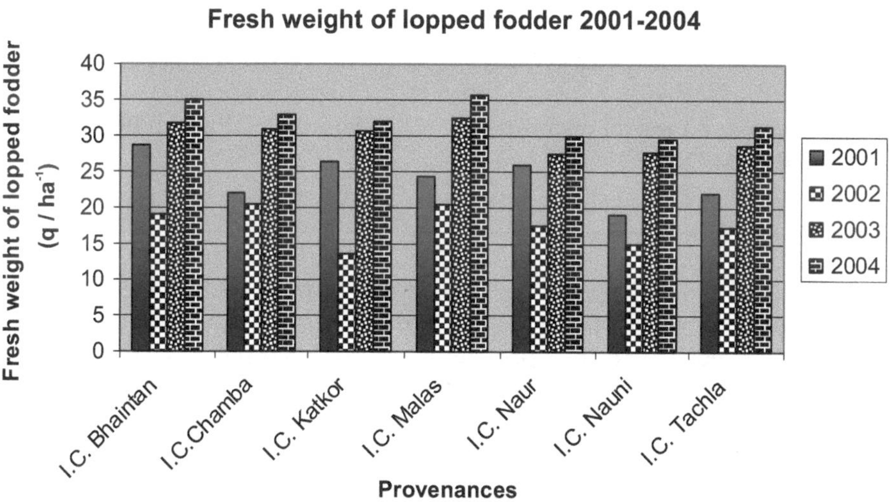

Fig. 1 : Fresh weight of lopped fodder in *Grewia optiva* during 2001-2004

planting were as follows : fresh weight of lopped fodder: 32.3 q ha^{-1}, dry weight of lopped fodder :14.0 q ha^{-1}, dry weight of debarked fuelwood 5.3 q ha^{-1}and dry weight of bark fibre : 2.6 q ha^{-1} (Table 6). The highest fresh weight of lopped fodder was recorded in I.C. Malas (35.6q ha^{-1}) closely followed by I.C. Bhaintan at 35.1q ha^{-1} (Figure 1).

Table 6 : Variability estimates of growth and productivity parameters in *Grewia optiva* after 10 years of planting at Dehradun

S. No.	Characters	Range	Mean	SE (m) ±	C.V.
(a) Growth parameters					
1.	Plant height (m),PH	5.3-6.2	5.6	0.21	6.3
2.	Collar diameter (cm),CD	11.1-14.4	13.1	0.78	10.2
3.	Diameter at breast height (cm),DBH	4.3-5.0	4.6	0.18	6.7
4.	No. of branches plant^{-1}, NB	18.0-26.0	21.3	0.65	5.3
5.	Crown length (m), CL	3.8-4.3	4.1	0.17	7.3
6.	Crown width (m), CW	2.8-3.3	2.9	0.08	4.9
(b) Phenological parameters					
1.	1000 fruit weight (g)	100.0-109.2	105.0	1.42	2.4
2.	Fruit production (g tree^{-1})	5.27-682	604	10.20	2.0
(c) Productivity parameters					
1.	Litter fall (t ha^{-1})	5.3-6.8	5.0	0.10	3.0
2.	Fresh weight of lopped fodder (q ha^{-1})	29.4-35.3	32.3	0.42	2.2
3.	Oven dried weight of lopped fodder (q ha^{-1})	12.5-15.4	14.0	0.23	2.8
4.	Oven dried weight of debarked fuel wood (q ha^{-1})	4.6-7.2	5.3	0.10	2.8
5.	Oven dried weight of bark fibre(q ha^{-1})	1.9-3.7	2.6	0.04	2.7

The fresh fodder weight of I.C. Malas (35.6 q ha^{-1}) was 1.21 times higher than the least productive provenance I.C. Nauni. Semwal *et al.* (2002) reported tree fodder productivity of 15.8 q ha^{-1} (3x3 m spacing) after 6 years while Singh *et al.* (2008) have reported the highest fresh fodder yield of 14 .0 q ha^{-1} (4x 4m spacing) under 75% lopping in 7th year after planting. Dry fodder yield ranged between 7.47- 14.02 q ha^{-1} as trial average over provenances for the years 2001 to 2004, respectively (Table 5) in comparison to the highest value of 8.0 q ha^{-1}

(Vishwanatham *et al*, 1999). Such differences could be primarily attributed to the site conditions and physico chemical properties of soil where these trials have been conducted as well as different germ plasm materials. Earlier studies have been conducted on gravelly flood plains and the marginal lands of Doon valley while the present studies have evaluated the maximum genetic potential of the provenances under good fertility conditions. This allowed the immediate selection of superior provenances and individuals within a provenance. Similarly, the dry weight of debarked fuelwood was the highest in I.C. Bhaintan at 7.14 ± 0.87 q ha^{-1} (Table 5) which was 1.81 times higher than the lowest productive provenance I.C. Tachla (3.95 q ha^{-1}). I.C. Bhaintan proved to be the best provenance for fibre extraction at 3.73 q ha^{-1} which was 2 times greater than I.C. Tachla.

The highest fresh and dry weight of barked and debarked fuelwood and dry weight of bark fibre were obtained in I.C.Bhaintan (27.8 and 8.1q ha^{-1}), respectively.

Leaf litter yield

The leaf litter yield averaged for 12 months was recorded highest in the provenance I.C. Malas (6.8 ton ha^{-1}) followed by I.C. Bhaintan (6.3ton ha^{-1}). The range of litter fall was 6.82 to 5.29 ton ha^{-1} for I.C. Malas and I.C. Nauni, respectively (Table 4).

The differences among provenances in respect of plant height, collar diameter, diameter at breast height, crown length and width were non-significant. Sehgal *et al* , 2003 also did not detect major differences for growth and nutritional parameters. The average plant height, collar diameter, diameter at breast height, crown length and crown width in final year (2004) were 5.6 m, 13.1 cm, 4.6cm, 4.1 m and 2.9 m, respectively (Table 4). Estimates of variances, coefficients of variability and genetic components in *Grewia optiva*, 10 years after planting at Dehradun (1995-2004) are presented in Table 7. It was indicated that genetic variation for growth parameters was negligible however, genetic variation for productivity parameters was substantial which leaves sufficient scope for selection of high yielding plant types (Mehta *et al*, 2007).

Phenological traits

The *Grewia optiva* provenances came to flowering after 5 years in 2000. The range of variation for fruit production was recorded from 527 to 682 in 2004 with an average of 603 fruits. Highest 1000 fruit weight was recorded in I.C. Naur and I.C. Nauni, respectively.

Correlation coefficients determined for different productivity parameters between years starting from 2001 (Table 6) indicated that fodder productivity in Ist year of lopping was poorly correlated with fodder productivity in subsequent years (2002, 2003 and 2004). However, fodder productivity levels of 2nd year of lopping were strongly correlated with 3rd and 4th year(p =0.01). The correlation coefficients were quite stronger for 3rd and 4th year (>0.90**) which indicated that the predictable comparative expression of genetic potential was noticeable after 2nd year of lopping than Ist year. Lopping yield of first year was not a true reflection of genetic potential of a provenance as it is clubbed with genotype x year interaction effects accumulated over 5 years.

The overall ranking of *Grewia optiva* provenances was worked out on the basis of 8 growth and phenological traits and 5 productivity parameters as overall rank sum index (ORSI) value. This index gives a single cumulative value with which provenances could be discriminated and the best provenances identified. ORSI index was the highest in case of I.C. Malas (0.21) followed by I.C. Bhaintan (0.20) thereby indicating that these are the best available provenances is respect of fodder productivity, fibre and fuelwood production. They also embody suitable ideo types and are recommended to be adapted in agroforestry systems. The next best provenance was I.C. Chamba with a cumulative score of 0.14. Further, the provenances I.C. Bhaintan and I.C. Malas have higher fecundity (more number of fruits) which makes them easy to propagate through seedlings.

Based on 10 year studies (1995-2004) it was clear that ample genetic variation existed among provenances for growth and productivity parameters. Three best provenances *viz.* I.C. Malas, I.C. Bhaintan and I.C. Chamba have been narrowed down which adequately meet agroforestry requirements. These provenances can be used as the basic stocks for developing quality planting materials in clonal orchards. Progeny trials can be initiated by selecting superior 'plus' type trees within each of these provenances and intra provenance variation exploited.

In continuation of the present study, 1200 saplings of these three best provenances were developed in a poly house during 2005-06. One year old saplings were subjected to multi-location testing at 4 sites in the North-Western Himalayas at diverse elevation in valley and hilly areas ranging from 515 m to 1750 m asl. on farmers' fields under different agroforestry systems. Significant provenance x location interactions have been detected in 2008 and 2009 and it will be possible to develop location specific technologies in respect of this agroforestry tree species in near future.

Studies on root parameters

Agroforestry tree species specially in rainfed Indian tropics and sub tropics have high value in respect of risk mitigation, catering domestic needs of fuel, fodder, fibre, timber fruit etc, soil fertility build up and conservation of natural resources and environment. Information on rooting pattern, above ground biomass, fertility build up and soil binding of perennial woody components particularly multipurpose tree species in agroforestry systems is very meager. Dadhwal *et al.* (2004) have studied the rooting behaviour of multipurpose tree species in agroforestry land use and fertility build up in degraded lands of north western Himalayas. Parameters studied included specific root length, biomass, volume, number, average diameter and horizontal and vertical spread of roots, soil binding, litter production, nutrient cycling and improvement in soil fertility in 11 years old agroforestry species. Since fine roots (<0.5 mm) mainly governs soil binding properties, it was observed that among MPTs *Grewia optiva* had the highest soil binding property followed by *Bauhinia purpurea* and *Eucalyptus* hybrid. Nutrient addition due to MPTs increased with age and decreased with radial distance from the tree base. Singh *et al.* (2000) studied the rooting behaviour of 4 agroforestry species viz., *Eucalyptus tereticornis, Leucaena leucocephala, Grewia optiva* and *Bauhinia purpurea.* They found that fast growing species *Eucalyptus tereticornis* and *Leucaena leucocephala* had relatively more aggressive root system than slow growing native species (*Grewia optiva* and *Bauhinia purpurea*). These species showed much less root density at 1 m depth compared to the former species therefore; these indigenous species are likely to be less competitive with crops in agroforestry systems than the fast growing species. *Grewia optiva* having shallow root system could be more effective for erosion control particularly on terrace risers while *Bauhinia purpurea* with deep root system and high root:shoot ratio could be more suitable for slope stabilization. Irrespective of tree species, mean annual leaf litterfall biomass showed a direct positive relationship with mean annual below ground fine root mass (R^2 = 0.97, $P<0.05$), when averaged across all sites. This information is useful to predict the mean annual fine root mass of the semi-arid tree species, as root production is time consuming, laborious and difficult to estimate (Jha and Mahapatra, 2009).

Root parameters were also studied in 3 years old saplings of high yielding provenances of Bhimal *viz.*, I.C. Bhaintan , I.C. Chamba and I.C. Malas on the basis of five plants with three replications (Dadhwal and Mehta, 2009). The entire root system of three provenances were excavated using skeleton method (Bohm,1979) One cubic meter pits were dug out and average root spread, average oven dry weight of above ground biomass and fine root estimations were made. Soil binding factor was worked out for different provenances. Soil binding factors of different provenances *viz.*, I.C. Bhaintan , I.C. Chamba and I.C. Malas showed a significant positive relationship with average oven dry weight of above ground biomass and specific root length. (Figures 2 a-f).

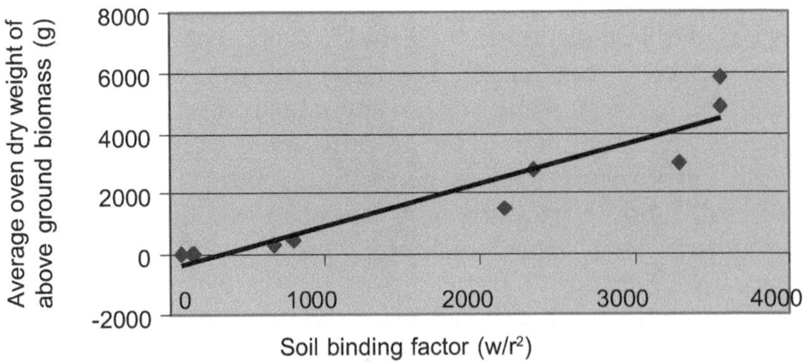

Fig. 2a : Regression of average oven dry weight of above ground biomass on soil binding factor of *Grewia optiva* (I.C. Bhaintan) y = 1.3832x-427.6 R^2 = 0.8917

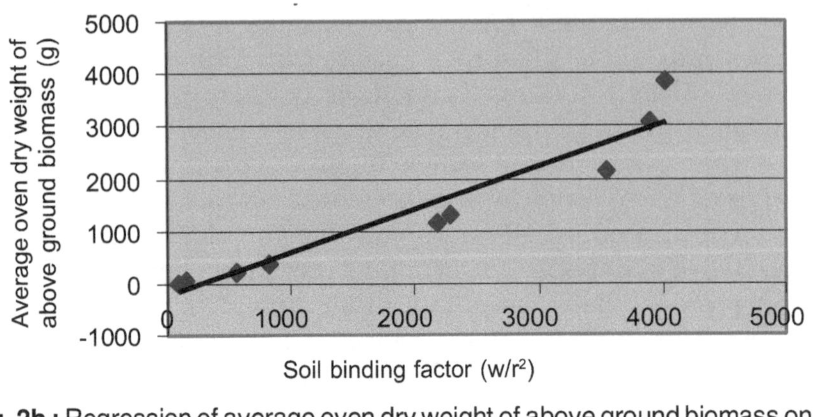

Fig. 2b : Regression of average oven dry weight of above ground biomass on soil binding factor of *Grewia optiva* (I.C. Chamba) y = 0.8219x-240.9 R^2 = 0.9281

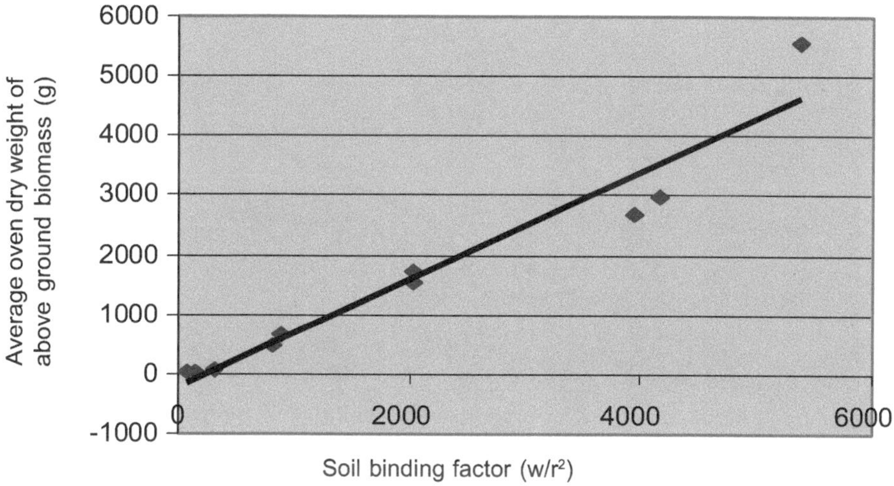

Fig. 2c : Regression of average oven dry weight of above ground biomass on soil binding factor of *Grewia optiva* (I.C. Malas) y = 0.9035x-212.9 R^2 = 0.9412

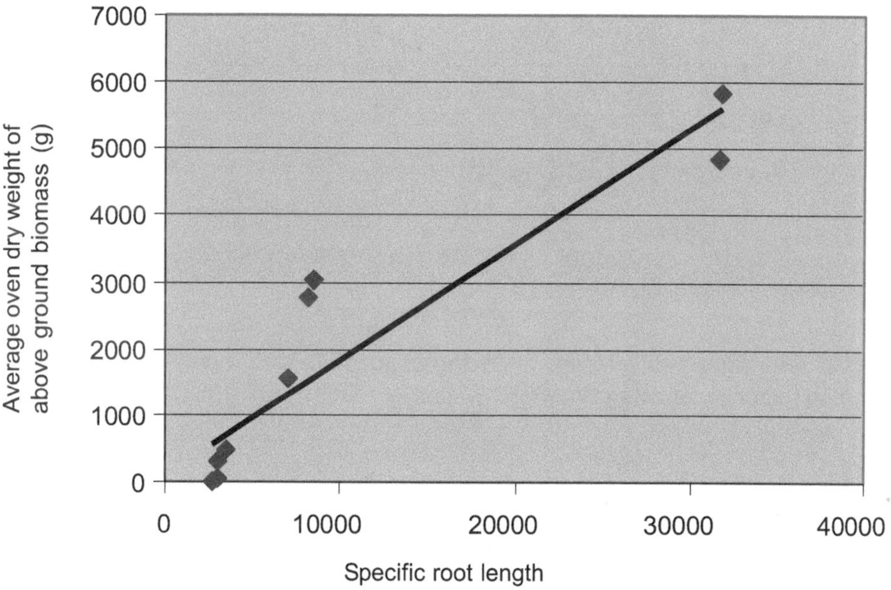

Fig. 2d : Regression of average oven dry weight of above ground biomass on soil binding factor of *Grewia optiva* (I.C. Bhaintan) y = 0.1734x-106.61 R^2 = 0.8664

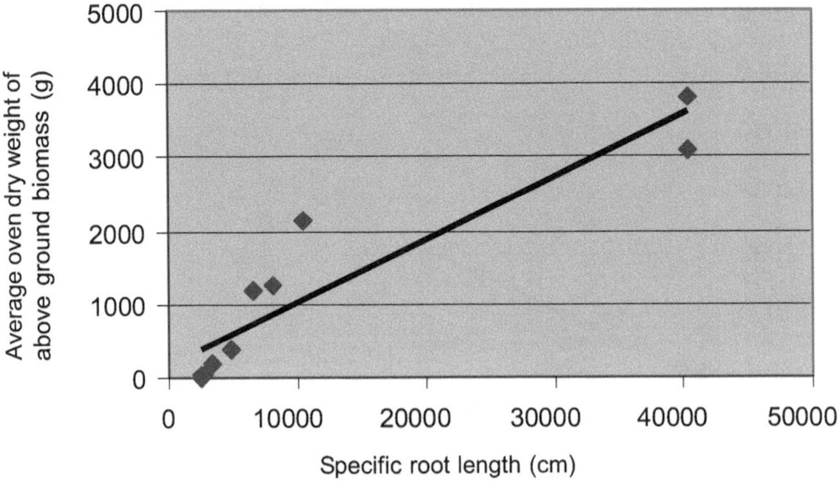

Fig. 2e : Regression of average oven dry weight of above ground biomass on soil binding factor of *Grewia optiva* (I.C. Chamba) y = 0.0854x + 173.11 R^2 = 0.8632

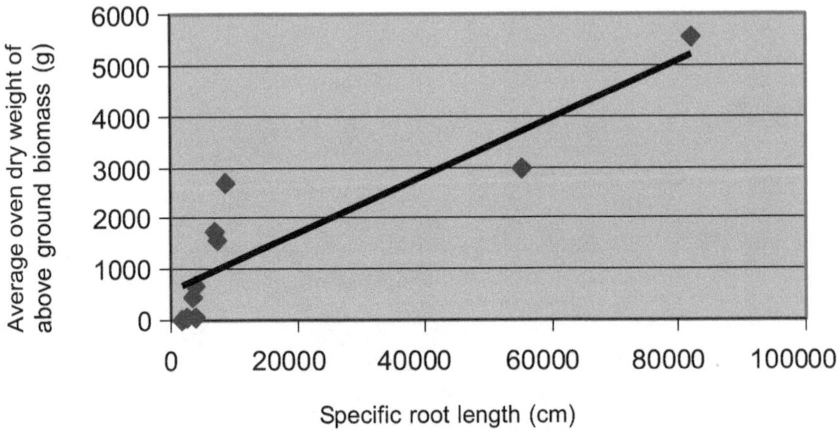

Fig. 2f : Regression of average oven dry weight of above ground biomass on soil binding factor of *Grewia optiva* (I.C. Malas) y = 0.0571x = 566.03 R^2 = 0.7987

Intercropping Studies of Bhimal (*Grewia optiva* vis a vis other Agroforestry Tree Species with Tranditional Crops

a. International status

The modern focus on sustainable agriculture and conservation of nature and landscape in Europe has increased the interest in silvo arable systems (Eichhorn *et al.*, 2006). Target regions within 32 European countries for silvo arable forestry including 5 commercial tree species (*Prunus avium*, *Juglans* sp., *Populus* sp, *Pinus pinea* and *Quercus ilex*). were identified and it was concluded that at least one of the species could grow productively in agroforestry context on 56% of the utilizable arable land while contributing simultaneously for protection of soil on 6 m ha , combat nitrogen leaching on 30 m ha and increase landscape diversity and hence potentially biodiversity on 42 m ha.

Green leaf manure (GLM) from leguminous trees into agroforestry systems can be substituted for inorganic N fertilizers to enhance crop growth and yield. Bayala *et al.* (2002) studied the effect of tree crown pruning in *Vitellaria paradosa* (Karite) and *Parkia biglobosa* (Nere) in agroforestry system with millets (*Pennisetum glaucum*) and the results showed that tree crown pruning had significant effect on millet production. The highest millet grain yield and total dry matter were produced under total pruned trees (5.07 and 2.03 q ha^{-1} yr^{-1}).

b. National status

India has been in the forefront of agroforestry research ever since organized research in agroforestry started about 30 years ago. Nearly 25 years of investment in agroforestry research have clearly demonstrated the potential of agroforestry in many parts of the country (Puri and Nair, 2004). Farmers in the north-western Himalayas generally follow *Grewia* based agroforestry, however scientific management of canopy and adequate number of plants per ha for optimum production from the system are yet to gain popularity among farmers. Minor millets like mandua (finger millet) and jhingora (barnyard millet) are the traditional cereal diet of hilly and tribal people and these crops are grown in association with agroforestry species like bhimal, mulberry, khirak and banj. The area and production of finger millet in Uttarakhand is 1,42,827 ha and 1,97,575 metric tonnes, respectively while that of barnyard millet is 68,033 ha and 89,443 metric tonnes respectively, next only to rice. These crops also provide palatable fodder to animals. They have profuse canopy and are helpful in resisting soil and water erosion on the steep slopes where they are cultivated. From 1980 to 1990 *Populus deltoides* has been planted commercially under agroforestry by farmers of north western India owing to its fast growth, high productivity (> 30 m^3 ha^{-1} yr^{-1},), deciduous nature in winter and intercropping with wide range of arable, vegetable, spices and fodder crops (NewMan, 1997). Among two predominant poplar based agroforestry systems, poplar + sugarcane and / or wheat were found to be most remunerative (Venkatesh *et al.*, 2009).

Khybri *et al.*, (1992) studied the tree crop interaction for 13 years from 1977 to 1990 under rainfed conditions of Dehradun valley. Bhimal, mulberry and eucalyptus hybrid were studied in rice (cv. Akashi) - wheat (cv. RR 21) crop rotation. All the trees were harvested in 1990, which showed depressing effect on crop yields. Eucalyptus had the maximum depressing effect on the crop yields till the first harvest and had the least effect thereafter. Mulberry affected rice most while wheat was affected by *Grewia optiva*. The depressing effect on an average varied from 28-34% depending on species. Competition for moisture, nutrients and light were the major factors which reduced the growth and yield of wheat in *Grewia* - wheat based agroforestry system (Kaushal *et al.*, 2005).

Tomar *et al.* (2004) evaluated the effect of planting direction and tree row distance of Bhimal for 13 years on field crops in marginal, bouldery riverbed degraded class V lands in Doon valley. Results indicated that above ground biomass (grain plus stalk) of crops of maize, wheat, and sesamum were 20.1, 4.8 and 31.3 % higher under *Grewia optiva* in NE-SW directions in comparison to SE-NW. Though the field crop yields were less under the tree –crop system but nutrient build up was higher under canopy than open area indicating improvement of degraded lands over time. Grewal *et al.* (1992) have carried out some agroforestry systems for possible adoption in place of high risk rainfed farming on land capability classes I to IV. The agrisilvihorticulture system integrating leucaena, lemon, papaya and turmeric on class I irrigated land provided sustainable mean net returns of Rs. 17066 against Rs. 7852 $ha^{-1}year^{-1}$ from double cropped agricultural system. The intercropping of cluster bean with leucaena gave the highest net returns of Rs. 3540 $ha^{-1}year^{-1}$ in agrisilvicultural system adopted on class II land. Singh *et al.* (2008) evaluated the management practices for a *Grewia optiva* - Hybrid Napier based silvipastoral system for rehabilitating old riverbed lands in the North-Western Himalayas and have recommended the adoption of coppicing and pollarding at 1-1.5 m height to facilitate the growth of annual crops. Similar recommendation was made by Verma *et al.* (2007). Hence, Grewia based agroforestry systems have great scopes of increasing farmers income on sustainable basis in this part of the country.

Table 7 : Variances, coefficients of variability and estimates of genetic components in *Grewia optiva*, 10 years after planting at Dehradun (1995-2004)

S. No.	Characters	Variances			Coefficients of variability		Genetic components	
		σ^2p	σ^2g	σ^2e	PCV	GCV	Heritability	Genetic advance (k=2.06)
(a) Growth parameters								
1.	Plant height (m)	0.17	0.04	0.13	7.28	3.67	0.25	0.22
2.	Collar diameter (cm)	2.49	0.68	1.81				
3.	Diameter at breast height (cm)	0.14	0.04	0.10	8.09	4.45	0.30	0.23
4.	No. of branches plant^{-1}	8.57	7.29	1.28				
5.	Crown length (m)	0.10	-	0.09	7.81	2.74	0.12	0.08
6.	Crown width (m)	0.04	0.02	0.02	7.12	5.17	0.53	0.23
(b) Phenological parameters								
1.	1000 fruit weight (g)	15.2	9.06	6.12	3.71	2.87	0.60	4.79
2.	Fruit production (g tree^{-1})	3765	3451	313	10.17	9.73	0.92	115.88
(c) Productivity parameters								
1.	Litter fall (t ha^{-1})	0.30	0.27	0.03	9.27	8.71	0.89	1.01
2.	Fresh weight of lopped fodder (q ha^{-1})	5.51	5.00	0.52	7.27	6.92	0.90	4.38
3.	Oven dried weight of lopped fodder (q ha^{-1})	1.50	1.34	0.16	8.74	8.27	0.89	2.26
4.	Oven dried weight of debarked fuel wood (q ha^{-1})	10925	10695	230	19.62	6.92	0.98	210.79
5.	Oven dried weight of bark fibre (q ha^{-1})	4091	4041	49.7	24.71	24.56	0.99	130.16

References

Bayala J., Z. Teklehaimanot and S.J. Ouedrago. (2002). Millet production under pruned tree crowns in a parkland system in Burkina Faso. *Agroforestry Systems*, 54: 203-214.

Bohm W. (1979). Excavation methods. Pp 5-19. In: Methods of Studying Root Systems. Spriger -Verlag, Berlin, Heidelberg.

Dadhwal, K.S. and H. Mehta (2009). Rooting behaviour of different provenances of Bhimal (unpublished), CSWCRTI, Dehradun

Dadhwal, K.S., P. Narain and S.K. Dhyani (1989). Agroforestry systems in the Garhwal Himalayas of India. *Agroforestry Systems*, 7: 213-225.

Dadhwal, K.S., V.P. S. Tomar and R.K. Dubey (2004). Rooting behaviour of multipurpose tree species in agroforestry land use and fertility build up in degraded lands of North West Himalayas. *The Ekologia*, 2 (1-2): 61-78.

Deb Roy, R; Shankaranarayan, K.A. and Pathak P.S. (1980). Fodder trees and their importance in India. *Indian Forester*, 106:306-311.

Eichhorn M.P., P. Paris, F. Herzog, L.D. Incoll, F. Liagre, K. Mantzanas, M. Mayus, G. Moren, V.P. Papanastasis, D.J. Pilbeam, A. Pisnelli and C. Dupraz. (2006). Silvo arable systems in Europe - Past, present and future prospects. *Agroforestry System*, 67:29-50.

Grewal S.S. , S.P. Mittal, S.Dayal and Y. Agnihotri (1992). Agroforestry systems for soil and water conservation and sustainable production from foothill areas of North India. *Agroforestry Systems*, 17(7):183-192.

Jha P. and K.P. Mahapatra (2009). Leaf litterfall, fine root production and turnover in four major tree species of the semi-arid region of India. *Plant and Soil*: Published on line.

Joshi P. and P. Narain (1992). Bhimal - a multipurpose tree for agroforestry. *Indian Farming*, 41(12): 14-15.

Kaushal R., K.S Verma and K.N. Singh (2005). Above and below ground interactions in *Grewia* -Wheat based agroforestry system. 295-303 pp. In : Short Rotation Forestry for Industrial and Rural Development (Eds. K.S.Verma , D.K.Khurana and Lars Christersson) ISTS Nauni, Solan India

Khybri M.L., R.K. Gupta, S.Ram and H.P.S. Tomar (1992). Crop yields of rice and wheat grown in rotation as intercrops with three species in the outer hills of Western Himalaya. *Agroforestry Systems*, 17: 193-204.

Mehta H., P.C. Tyagi and K.S. Dadhwal (2007). Genetic variation for growth and productivity parameters in a collection of Bhimal (*Grewia optiva* J.R. Drummond ex Burret) provenances collected from north western Himalayas. *Range Mgmt. & Agroforestry*, 28(2), 323-325.

Negi, S.S. (1977). Fodder trees in Himachal Pradesh. *Indian forester*, 103: 616-622.

NewMan, S.M.(1997). Poplar agroforestry in India. *Forest Ecology and Management*, 90:13-17.

Puri S. and P.K.R. Nair. (2004). Agroforestry research for development in India: 25 years of experiences of a national programme. *Agroforestry Systems*, 61: 437-452.

Raghvan, G.V. (1990). *Availability and use of shrubs and tree fodder in India.* In: Shrubs and Tree Fodder for farm animals (ed. C. Devendra) pp. 198-210,. International Development Research Centre, Ottawa Canada.

Rekib, A., V.S. Upadhyay, and P.S. Pathak (1991). Trees for fodder. *Indian Farming*, 41:56-61.

Sehgal R.N., A. Rathore and S.K. Chauhan (2003). Divergence studies in selected genotypes of *Grewia optiva* . *Indian J. Agroforestry* 5 (1): 99-102.

Semwal R.L., R.K Maikhuri., K.S Rao., K. Singh and K.G. Saxena (2002). Crop productivity under differently lopped canopies of multipurpose trees in Central Himalaya, India. *Agroforestry Systems*, 56: 57-62.

Semwal, R.L. and R.K. Maikhuri (1996). Agro-ecosystem analysis of Garhwal Himalayas. *Biological Agriculture and Horticulture* :13.

Singh C., A. Raizada, M.K., Vishwanatham, S.C. Mohan (2008). Evaluation of management practices for a *Grewia optiva* - Hybrid Napier based silvi-pastoral system for rehabilitating old riverbed lands in the North-Western Himalayas. *Int. J. Ecol. and Environ. Management* 34 (4): 319-327.

Singh R.K., P. Narain, S.K. Dhyani and J.S. Samra. (2000). The rooting behaviour of four agroforestry species in the western Himalayan region. *Journal of Tropical Forest Science*, 12(2): 202-220.

Singh, R.V. (1982). *Fodder trees of India.* Oxford and IBH Publishers, New Delhi.

Tomar, V.P.S., K.S. Dadhwal, B.N. Ghosh, H.P.S. Tomar and D.Panwar (2004) Directional effects of multipurpose tree species (MPTs) on crops in degraded lands of N-W Uttarachal Himalayas. *J.Natcon*,16(2):267-276.

Venkatesh A., R.K. Tiwari, R.Singh, K Kareemula, A.Kumar, O.P. Chaturvedi and S.K. Dhyani.(2009). *Biometrics and economics of poplar based agroforestry systems in northern India.* In: Agroforesry: Natural Resource Sustainability, Livelihood and Climate Moderation (Eds. O.P. Chaturvedi *et al*,) pp. 661-674.Satish Serial Publishing House, Delhi.

Verma K.S, D.R. Bhardwaj and K.Chand.(2007). *Agroforestry Systems in Himachal Pradesh*, pp. 67-93. In: Agroforestry Systems and Practices (Eds. Puri S. and P. Panwar) New India Publishing Agency Pitampura, New Delhi 110088.

Viswanatham, M.K., J.S. Samra, and A.R. Sharma (1999). Biomass production of trees and grasses in a silvipasture system on marginal lands of Doon valley of north-west India. I. Performance of tree species. *Agroforestry systems*, 46 (2):181-196.

Chapter-11

Tree Based Farming Systems for Paonta Valley of Himachal Pradesh

V.K. CHAUHAN, A.K. JOSHI AND R.S. RANA
Horticultural Regional Research Station, Dhaulakuan, Sirmour-173 001 (HP)

ABSTRACT

Survey and research trial conducted for the selection of promising tree based farming system in Paonta valley region of H.P. have been summarized in the manuscript. Tree based farming innovations with intercropping, their planting technology and geometry around field/ farm, pure plantation , block plantation of trees grown with agricultural / fruit crops for better productivity per unit of land have been suggested for improving socio-economic condition of the regional farmers.

INTRODUCTION

Tree based farming system / agroforestry is a dynamic, ecologically based, natural resources management system that, through the integration of trees on farms and in the agricultural landscape, diversifies and sustains production for increased social, economic and environmental benefits for land users at all levels (ICRAF, 2006). Growing woody perennial with agricultural crops has variable meaning around the globe to produce diverse products per unit of land which not only benefit farmers but also generates employment benefiting sizeable masses in related economic activities like nursery raising, plantation, their management, felling, logging operation, transportation, sawing and conversion to desired product and their marketing. Crops grown underneath tree are of secondary consideration and form a small part of gross income. These crops compliment wood production and boost early cash inflow. Survey and research trials on poplar in agroforestry at Horticultural Regional Research Centre, Dhaulakuan (H.P) revealed its deciduous nature, straight clean bole, narrow and sparse canopy, multipurpose use of wood, fast growth ranging from 15 to 23 $m^3 ha^{-1}yr^{-1}$ on a short rotation of 6-7years and high economic returns makes poplar versatile and most ideal tree species for tree based farming system in Paonta Valley regions of Shivalik of lower western Himalayas.

Tree based farming system have one commonality that is deliberate use of trees and shrubs (Woody components) in order to achieve higher productivity, sustainability and diversity of output from same land use system. That makes agroforestry a distinct discipline obviously related to and dependent on the body of existing knowledge in conventional agriculture and forestry discipline. The science of agroforestry is, thus, a matter of finding and developing the relevant combination of woody and non-woody components in relation to the land user's problems, aspirations and potentials. Also to develop special planting arrangement and management practices which minimize the competitive interactions between the components and to maximize the productive and service functions of the trees (Lundgren,1982). To qualify as agroforestry, there must be both ecological and economical interactions between the woody and non-woody components. Agroforestry approach aims at not only the production of variety of products i.e food, fodder, fuelwood and timber but also ensure the conservation and amelioration of producing environment. It holds most apparent potential in marginal and fragile areas and in resource limiting small land holder system where monoculture, agriculture or forestry may not be feasible. Because, the leaf litter of deciduous trees when used as mulch not only protect the soil from the impact of the rain but also improve soil structure, reduces evaporation and increases infiltration (Young, 1986).

The economic fortunes of mountains are inextricably bound with the state of their natural resources like perennial water supply of river system, wasteland, grasslands and forest. These natural resources ensure the grazing for cattle, small wood for agriculture implements, and in particular, firewood to ensure the cattle dung for manure to boost food crop production. The mountain inhabitants, by and large, are faced with the widening gaps between demand and supply of basic needs i. e. food, fruit, fodder, fuel, timber and related land based products causing increasing problems of sustenance due to lack of suitable tree based farming system developed in isolation over time without considering the diverse needs of hill ecosystems. For example, the potential of traditional tree- crop- livestock production system which have sustained the mountain agriculture since ages through their inherent capabilities to sustain life and also to protect the producing environment has not been realized while delivering the current research and development interventions. Blending of agriculture with forestry is necessary for the well being of mankind (Gill and Lal, 2000). The complimentary or supplementary effect of integrating trees with crops depend upon the kind of the associated crop. Kumar, 2006 took 67 cases of tree and crop association for assessing the positive and negative effect of trees on crops. The study indicates that crops such as upland rice, ginger (*Zingiber officinale*), and *Kaempferia galanga* showed higher productivity in certain agroforestry combinations (over sole crops), while fodder plants and many other grain crops showed relatively lower yields. That is, of the 67 cases (48 species combinations, some in more than one management situations), 16 showed positive effects, 12 depicted neutral effects, and another 39 illustrated negative trends. This chapter present the broader view of suitable tree based farming system for Paonta Valley region of Shivalik for improving the socio-economic condition of farmers and to serve as a tool for further development and planning.

Study Area

The state of Himachal Pradesh is located from 30° 22' 44" to 33° 12' 40" N latitude and 75° 45' 55" to 79° 04' 20" E longitude covering a total area of 5.57 million ha. Geographically, the state is mountain locked and the terrain is highly rugged. There are four major physiographic divisions- middle or central Himalayas comprising Kinnaur, Kullu, Lahaul- Spiti and Pangi areas, the lesser Himalayas of Dhauladhar ranges, Shimla hills and Karsog ranges, the Shivaliks of Sirmour, Bilaspur, Solan and Hamirpur areas and the fourth is river terraces or flood plains of Bahl valley, Kullu, Poanta, Una and Indora areas. The variation in altitude is from 350 to 6975 metres. The state represents much diversity in land use systems due to topographic variation. The Sirmour district having a geographical area of 2,24,752 ha out of which 41,626 ha (18.52%) is net cultivated area. Sirmour is one of the most backward district of Himachal Pradesh owing to its tough topography, illiteracy, small land holdings and traditional farming systems. The district is divided into six blocks i. e. Pacchad , Rajgarh, Nahan, Sangrah, Shillai and Poanta Sahib. Some parts of the district like Poanta Sahib have achieved astonishing progress in agriculture and industrial fronts mainly due to cereal cropping pattern of wheat, maize, ginger and sugarcane with bunds/periphery/block plantation of poplar, eucalyptus, toon , shisham, beul and mulberry deliberately grown for fuel , fodder and small timber purposes and also because of their easy marketing option in nearby states of Haryana and Uttarakhand. The district comprises of two distinct climatic zones. One comprising of plain areas and low valleys so called the foothill zones, upto an altitude of 915 m a. m. s. l. having sub-tropical conditions (Poanta valley). The other with sub-temperate sub-humid, called the mid hill zone upto an altitude of 915-1524 m a.m.s.l.

Methodology

Selection of suitable tree based farming system for a particular region is an important and complex subject, upon which its success depends. Tree-crop interaction is a complex phenomenon in which the ecological requirements of crop plants change with tree density and geometry. Increase in tree size induces significant changes in environmental parameters and affect agricultural crop production. At later stages of tree growth, these changes are much pronounced and significantly reduce the yield of most agricultural crops, though the over all productivity of agroforestry system remains higher when compared with the sole cultivation of crops or tree component (Harsh and Tewari, 1993). The successful research finding for selection of suitable tree and their planting technology to be grown with traditional agricultural crops evaluated at this centre are enlisted hereunder:

Planting Around Field / Farm

Trees are planted in a single row around the periphery of field/ farm at 2 to 3m distances. On the periphery of one hectare land, 200 plants of poplar can be planted at 2m distance which produces on an average 3 quintals of bole wood/tree in 6 to7 years with proper management practices. Considering an average price of Rs.400 per quintals of bole wood as it prevails in North Indian market (Jagadhari and Yamunanagar), the tree grown on periphery of one ha. land can yield about Rs. 2.4 lacs. This is the additional income which the farmer obtain from same unit of land besides cultivating traditional agricultural crops such as wheat, maize and sugarcane. This planting system is capable to yield an annual income of about Rs. one lakh, without any extra management practices of tree component. Tree planted around the field/farm not only showed non- significant effect on economic yield of associated crops but also improved the favourable microclimatic conditions responsible for complementary interaction among bio-physical factors. Poplar cultivation in agroforestry provides additional income through sale of tree produce, improves soil conditions by addition of organic matter and prevent future nutrient decline (Singhal and Panwar, 1991).

Pure Plantation

Poplar planted as pure crop at 3x3m distance, accommodate about 1111 plants per ha (Table 1). The plantation so raised with optimum management practices can provide an income of about Rs.13.33 lacs per ha on 6 to 7 years of rotation. The system is capable to yield an annual income of Rs. two lacs and proves good for absentee landlord. The farmers are advised not to grow the intercrops under the system. However, the perennial fodder grasses can be grown upto first two year of plantation to yield indirect benefit to the growers.

Table 1 : Tree based innovations developed for Paonta valley region of Shivalik of lower western Himalayas

Particular of innovations	Plants ha^{-1}	Yield q ha^{-1}	Age in years	Returns Rs. ha^{-1} (Lacs)	Av. annual returns (Rs. lacs)
Mango orchard alone	100	200	10	2.00	0.2
Kinnow- mandarin alone	400	520	10	5.25	0.52
Papaya alone	2500	1500	2	7.5	3.7
Strawberry alone	60000	120	1	7.0	7.0
Poplar alone	1111	3333	7	13.13	1.8
Wheat/maize alone	--	40,30	1	0.8	0.8
Sugarcane alone	--	450	1	0.9	0.9
Mango + Poplar on periphery	100+200	200 ,600	10,7	4.4	0.5

Mango + Poplar planted in block simultaneously	100,500	150 ,1500	10,7	7.5	1.0
Mango+wheat/maize+ poplar on periphery	100+200	200 ,600	10,7	4.4	1.14
Mango+ Poplar (block plantation) + wheat/maize	100 ,500	150, 1500	10,7	7.5	1.4
Poplar + sugarcane	400	1200	7	4.8	1.4
Kinnow +poplar on periphery	400 ,200	520,600	8,7	7.6	0.9
Poplar +turmeric/Ginger	400	1200	7	6.2	1.7
Poplar+wheat/maize	400	1200	7	4.8	1.5

Photo 1 : Poplar+ wheat

Photo 2 : Kinnow+Poplar+Wheat

Block Plantation with Intercrops

Trees are planted in the entire field at various planting geometry in order to create wider inter- row distances with decrease in within-row distances for creating space for maintaining optimum yield of agricultural crops. The planting distances can be 8x3 m or 6x4 m keeping in view the nature of tree and type of intercrops to be grown with tree (Photo 1). This planting technology is most suitable for small and marginal farmer who can undertake the intensive agronomic/ silvicultural practices to boost the productivity of crop and tree component. Manpower of farmer and his family member is most efficiently used for undertaking day to day operation in this system. Sugarcane, wheat, maize, ginger, turmeric and promising medicinal plants like ghee-kanwar, tulsi, akarkara and aromatic plants can be successfully grown under this system. This tree based farming system is capable to yield about Rs. 1.5 lacs per annum to the grower per hectare of land when grown in proper consultation with the agroforester.

Planting with Horticultural Crops

Planting tree with mango, citrus, papaya and strawberry is an economically sound, risk free entrepreneur for small and marginal farmers for increasing their income per unit of land. The promising planting system are as below :

i) Planting around orchard

Trees planted at 2 to 3 m distances around the orchard, protect fruit plant from wind storm and reduces the chances of flower drop and enhance fruit yield considerably by improving favourable microclimate. Besides the trees grown around the orchard provide additional income to the growers.

ii) Planting tree in the centre of four fruit trees

Fruit trees are planted at recommended spacing for better growth and fruit yield. Studies conducted by Chauhan *et al* (1997) on growth and productivity evaluation of an agrihortisilviculture system revealed that subabool trees can be interplanted beneficially to cover 45 per cent area in kinnow- mandarin orchard and the system is capable to provide 25 to 31 per cent higher monetary gains in comparison to planting of kinnow mandarin orchard alone (control plot treatment).

iii) Planting poplar either in between the plant to plant or row to row of fruit trees

Poplar plants are deliberately planted only in one direction of mango/ citrus rows in a orchard (i.e either in between the plant to plant or in between the row to row) in order to facilitate field ploughing and availability of sun light (Photo 2). Agricultural crop can be successfully grown in the inter spaces of woody perennial. Growers obtain produce from three component i.e horticulture, forestry and agriculture and that is why it is popularly known as three dimensional farming which is the utmost need of the day. This planting system is most successful in young or in newly planted orchard because commercial fruit yield starts after six or seven years and within this period poplar tree can be harvested and sold, making rooms for other component. There is no adverse effect of poplar on establishment and subsequent growth of fruit trees. In addition poplar trees nurses and shelter the juvenile orchards. This farming system comprises wide range of subjects in which the input- output ratio is quite high per unit of land. Growers are advised to consult the expert before taking up these tree based farming system.

Future Needs for Tree Base Farming Systems

The rapid increase in human population, industrialization and urbanization have resulted into scarcity of land. Growing scarcity of land results in land use intensification i.e high external input agriculture but strengthens the trend towards multiple cropping and interaction between the various production processes i.e. low external input sustainable agriculture (LISA) which is an intimate form of present day tree base land use system. Planning and developing tree based farming system is a suitable and adoptable forms of land use option, therefore, should take into account the following issues :

1. Planning, implementation and evaluation of relevant tree based farming system development programmes must be a joint venture of a multidisciplinary team of scientists.

2. For achieving long term gains and stability in the production potentials of resources available, some changes in R and D of agroforestry are required such as under:

   a) *Education*: Agroforestry should form an integral part of study at graduate level in agriculture, horticulture and forestry disciplines. At post graduate level, agroforestry education should be a specialized one at par with other discipline.

   b) *Research* : Agroforestry research should be supported with multi disciplinary team of scientist with clear cut frame of policy alongwith working manpower and funds.

   c) *Development* : Agroforestry should form as an integral and essential component along with separate resource allocation at ministerial and institutional level.

   d) *Policy frame work* : Agroforestry despite having immense potential of sustained economic upliftment of small and marginal farmers, does not have a clear policy framework. Therefore, well funded policy form an important pre-requisite for overall development of the subject, which shows commitment, intention of the Government in shaping progress of the sector over a long period of time. Absence of policy results in haphazard treatment of subject which ultimately leads to negative growth of the sector.

References

Chauhan, V.K., V.K. Mishra and P.K. Khosla (1997). Growth and productivity evaluation of an agri-horti-silviculture system. *J. Tree Sciences*. 1- 6(2) : 67-74.

Gill A.S. and B. Lal (2000). Agro-forestry system for rainfed black soils. *Indian Forester* 126 (4): 367-375.

Harsh, L.N. and J.C. Tewari (1993). Tree-crop interactions in agroforestry practices. In :*Agroforestry For Rural Needs*, ed .Khurana, D. K. and Khosla, P. K.; ISTS, Nauni- Solan (H P): 535-541.

ICRAF, (2006). World Agroforestry Centre, Southeast Asia web site. (http://www.worldagroforestrycentre.org/sea).

Kumar, B.M. (2006). Agroforestry: the new old paradigm for Asian food security. *Journal of Tropical Agriculture* 44 (1-2): 1-14.

Lundgren, B. (1982). Introduction. *Agroforestry System*, 1: 3-6.

Singhal R.M. and B.P.S. Panwar (1991). A study of cropping pattern of poplar based agroforestry system in North- Western Uttar Pradesh. *Van Vigyan*. 29 (3) : 191.

Young A. (1986). Effect of trees on soils. ICRAF *Reprint* No. 31. Nairobi, Kenya.

Chapter-12

Aonla Based Agroforestry Systems for Soil and Water Conservation in Shivaliks

RAM PRASAD, R.P. YADAV AND PANKAJ PANWAR
Central Soil and Water Conservation Research and Training Institute
Research Station, Chandigarh- 160 019

ABSTRACT

A long term study was conducted to evaluate different aonla based silvipastoral systems for compatibility, productivity and conservation effects. Three grass species viz. dholu, hybrid napier and perennial arhar were evaluated for 14 years for degraded lands of Shivaliks. Hybrid napier because of its vigour growth adversely affected the growth of aonla but effect on yield turned non-significant. Arhar could survive only during pre-bearing state. Studies on soil losses and runoff for decades showed that *Embilica officinalis* + *Chrysopogon fulvus* was the best silvipastoral system for soil conservation. This system saved rainwater and soil in a range of 4.88 to 30.66 cm and 862-2818 kg ha^{-1} respectively over sole plantation. The combination of *Embilica officinalis* + *Chrysopogon fulvus* proved to be the most effective silvipastoral system for reclamation of degraded lands and improving economic status of people in Shivalik foot hills.

Introduction

Shivalik hills are the lowest and outer most ranges of the Himalayas running parallel to Himalayan ranges from Jammu and Kashmir to

Uttrakhand in India. Major physiographic units are hills, piedmont plains, rivers/choes and flood plains.The average rainfall for the region works out to be 1100 mm. Shivaliks are constrained due to land degradation from water erosion, due to burgeoning human and cattle population and high intensity rainfall resulting in low input use efficiency and loss of biodiversity. As a result Shivalik region has been identified as one of the most degraded rainfed agro-ecosystem of the country and is included in priority areas for watershed development. Moderate to severely degraded lands can be brought under production by adopting scientifically based alternative land use system which besides reversing degradation through soil and water conservation can also improve the economic condition of the farmers. Hortipasture or agrihorticulture has been found suitable for this region by many workers.

Aonla based Hortipasture System

Aonla (*Embilica officinalis*) known as Indian Gooseberry is an important fruit of great commercial and medicinal value. It belongs to family Euphorbiaceae. It is known for rejuvenating and revitalizing effect on human body being rich source of vitamin C and protects against oxidation.

Aonla is a subtropical fruit and is well adapted to varying edaphic and climatic conditions. Considering its adaptability attributes its cultivation in degraded lands to check soil erosion in Shivalik region as one of the alternative land use system was studied on long term basis for about 14 years. The results of this long term study on soil and water conservation is provided in following heads.

Experimental Lay Out

Aonla var Chakaiya based hortipasture system was established in 1991-92. The study consisted of four treatments viz. T_1 – sole *Embilica officinalis*, T_2: *Embilica officinalis* + *Chrysopogon fulvus* (Dholu grass), T_3: *Embilica officinalis* + *Pennisetum purpureum* (Hybrid Napier grass var. NB-21) and T_4: *Embilica officinalis* + *Cajanus cajan* (Perennial Pigeonpea var. ICP-8094) replicated three times having nine plants in each replication with a plot size of 18 m x 12 m. Aonla were planted in 1991 at a spacing of 6 m x 4 m in experimental farm of CSWCRTI, Research Centre- Chandigarh at Mansadevi, Panchkula.

Fruit and Intercrop Yield in the System

The study conducted over a period of 14 years revealed that effect of intercrops on growth of aonla depends upon their competitiveness for natural resources and growth characteristics of intercrops. The parameters of growth characteristics of aonla *viz.* basal area, gbh, height and canopy spread were affected similarly by intercrops but the extent of magnitude varied. Hybrid napier cultivation caused most significant reduction in all the aonla growth characteristics which is due to its vigorous growth. Cultivation of dholu failed to cause any reduction in majority of the growth characteristic of aonla. Thus on the basis of long term study it can be inferred that cultivation of *Chrysopogon fulvus* was most compatible between the two grasses. *Cajanus cajan* (arhar) is compatible in initial pre-bearing stage as it is fertility restorer and its effect on growth of aonla is synergetic in post bearing stages.

Fruit yield under various treatments are given in table 1. However, it was found that over a period of nine years the fruit yield among different treatment was found to be non-significant.

Table 1 : Fruit yield of aonla (Kg plant^{-1}) under various Hortipastoral systems

Treatment	Year									
	1995	1996	1997	1998	1999	2000	2001	2002	2003	2004
Aonla pure	27	20	29	78	75	94	119	99	109	139
Aonla + C. *fulvus*	31	23	21	67	68	94	90	105	124	152
Aonla + *P. purpureum*	20	16	11	33	54	76	63	92	92	122
Aonla + C. *Cajan*	28	21	34	55	86	73	79	105	128	132
CD (P=0.05)	NS	NS	10	NS	NS	NS	NS	NS	NS	NS

Dry matter yield (qha^{-1}) of different intercrops in post bearing years is provided in table 2. It was found that the reduction in dry matter production of dholu grass was from 11 per cent to 73 per cent over pure grass. Whereas the reduction in yield of P. *purpureum* in agroforestry system ranged from 12 per cent to 41 per cent over pure grass (Prasad *et al.,* 2007).

Table 2 : Dry matter yield (q ha^{-1}) of different intercrops under aonla.

Treatments	Yield (q ha^{-1})						
	1997	1998	1999	2000	2001	2002	2003
Pure *C. fulvus*	58.7	46.79	37.83	44.29	32.40	49.51	43.12
Aonla + C. *fulvus*	52.50	38.50	31.90	30.25	28.70	19.81	11.55
Reduction in yield (%)	11	18	16	32	11	60	73
Pure *P. Purpureum*	189.38	219.40	213.35	214.10	141.57	183.72	159.38
Aonla + *P. purpureum*	136.30	128.60	125.9	118.60	124.20	121.78	118.64
Reduction in yield (%)	28	41	41	45	12	34	26

Runoff and Sediment Loss under Aonla based Landuse Systems

Runoff

The runoff varied from 10.3 to 46 % under sole aonla (T_1) during 1993 to 2003 depending upon rainfall, its intensity and distribution and canopy development (Table 3). Whereas, runoff ranged from 0.5 to 19.0, 3.3 to 32.8 and 2.0 to 39.3 % with its overall average of 8.0, 13.1 and 18.6 % for aonla + *Chrysopogon* (T_2), aonla + hybrid napier (T_3) and aonla + perennial pigeon pea (T_4). Thus *Chrysopogon fulvus* proved to be best intercrop in aonla for controlling runoff followed by napier and pigeonpea. Runoff increased during post- bearing years by 26, 64, 83 and 62 % in T_1, T_2, T_3 and T_4 treatments, respectively (Yadav *et al.*, 2005).

Sediment losses

Soil loss under silvipastoral system followed the trend of runoff and varied in different years. Soil loss ranged from 696 to 7864 kg ha^{-1} during 1993-2003 under sole aonla (Table 4) Cultivation of *Chrysopogon*, napier and pigeon pea with aonla produced sediment losses of 170 to 1843, 547 to 4658 and 576 to 5812 kg ha^{-1} , respectively. The effectiveness of different intercrops was in the order of : *Chrysopogon*>napier>pigeonpea with 81, 56 and 25 % reduction in average sediment loss over sole aonla, respectively.

Table 3 : Runoff produced under different Hortipastoral systems

Year	Per cent runoff			
	T_1	T_2	T_3	T_4
		Pre bearing stage		
1993	10.3	2.9	3.3	2.0
1994	16.9	4.5	6.3	4.0
1995	38.5	12.7	21.4	33.8
1996	14.9	3.6	4.9	15.3
Average	20.2	5.9	9.0	13.8
		Post bearing stage		
1999	14.2	0.5	5.6	14.2
2000	25.2	9.1	17.3	21.5
2001	24.2	13.3	15.7	23.1
2002	49.0	19.0	32.8	39.3
2003	17.5	6.5	10.9	14.0
Average	25.4	9.7	16.5	22.4
Over all average	23.0	8.0	13.1	18.6

Table 4 : Sediment loss under different Hortipastoral systems

Year	Sediment losses (kg ha^{-1})			
	T_1	T_2	T_3	T_4
		Pre bearing stage		
1993	2565	420	547	576
1994	696	573	665	696
1995	7864	1843	4658	5812
1996	4222	745	1878	3167
Average	3837	895	1937	2563
		Post bearing stage		
1999	3176	170	910	2878
2000	1310	307	776	1049
2001	3119	550	1683	2986
2002	7119	856	1971	5100
2003	1085	324	640	1138
Average	3162	441	1196	2630
Over all average	3462	643	1525	2600

Resource Conservation through Agroforestry

Conservation of soil and rainwater by intercropping of different grasses was estimated. The maximum retention of soil was by *Chrysopogon* (2818 kg) followed by napier (1936 kg) and pigeonpea (862 kg). Reduction of runoff resulted in higher infiltration of rainwater into soil profile. The water conserved ranged from 5.16 to 30.66, 4.88 to 20.32 and negligible to 14.13 cm annually under different systems during 1993 to 2003. The average annual water conservation was highest with *Chrysopogon* followed by napier and pigeonpea. These systems conserved 15, 10 and 4 cm water under *Chrysopogon*, napier and pigeonpea, respectively over and above the sole aonla (Table 5).

Table 5 : Soil and water conserved under different Hortipastoral systems.

Year	Soil (kg ha^{-1})			Rain water (cm)		
	Ch	Na	Ar	Ch	Na	Ar
			Pre-bearing stage			
1993	2145	2018	1989	5.16	5.88	5.79
1994	123	31	0	13.52	11.63	14.13
1995	6021	3206	2052	30.66	20.32	5.62
1996	3477	2344	1055	12.02	10.59	(-)0.46
Average	2942	1900	1274	15.34	12.11	6.27
			Post-bearing stage			
1999	3006	2266	298	11.47	7.22	0.03
2000	1003	534	261	14.55	7.08	3.30
2001	2569	1436	133	10.70	8.28	1.13
2002	6269	5148	2019	25.49	12.46	6.28
2003	761	445	(-)53	12.86	7.79	4.15
Average	2722	1966	532	15.01	8.57	2.98
Over all average	2812	1936	862	15.16	10.04	4.44

Ch: *Chrysopogon fulvus*, Na: Napier, Ar: Pigeonpea.

References

Prasad, Ram, R.P. Yadav, R.K. Aggarwal, Y. Agnihotri, S.P. Mittal, A.K. Tiwari and J.S. Samra (2007). Prospects of aonla cultivation on degraded Shivalik lands. Tech. Bulletin no. T-55/C-13. Central Soil and Water Conservation Research and Training Institute, Research Centre, Chandigarh (U.T) p 133.

Yadav, R.P, Ram Prasad, R.K. Aggarwal, Y. Agnihotri and J. S Samra (2005). Aonla (*Embilica officinalis*) based silvi-pastoral systems for soil and water conservation in degraded lands of Shivalik foothills. *Indian J. Soil Cons.*, 33(3):207-212.

Chapter-13

Role of Agroforestry in Soil Conservation and Soil Quality

R.P. YADAV, S.K. YADAV, PANKAJ PANWAR AND S.L. ARYA
Central Soil and Water Conservation Research and Training Institute
Research Centre, Chandigarh–160 019

ABSTRACT

Tree component of agroforestry systems produce large biomass of varying nutrient concentration to serve as alternative to inorganic fertilizers under integrated nutrient management or organic farming. Additionally, agroforestry systems proved effective in conserving soil and rainfall and brought down soil loss up to 0.2 t ha^{-1} and runoff up to 1.8 per cent of rainfall. Introduction of grasses in aonla conserved up to 31 cm of rainwater, 28 q ha^{-1} of soil, 11 kg ha^{-1} of N and 2kg ha^{-1} of K_2O. Conservation coupled with ameliorative capacity of agroforestry systems helped in building organic carbon and nutrients and improving soil physical, chemical and biological properties.

Introduction

To sustain the current level of productivity, 50 per cent of nutrient requirement must be met from source other than inorganic fertilizers. Leaf litter in range of 0.12 to 31.8 q ha^{-1} is being supplied annually from the different multipurpose tree species (Singh *et al.*, 1998). Although nutrient concentration in the leaf litter under various agroforestry land use systems is far lesser than inorganic fertilizers, the bulk addition can improve nutrient composition of soil for all the essential plant nutrients. Their role to serve as alternate source for nutrients under integrated nutrient management for Shivalik has been highlighted by Aggarwal *et al.* (2000). Introduction of forestry or fruit species in place of raising traditional field crops or intercropping trees with field crops have proved to be helpful in restoring and sustaining productivity of soils. Eco-friendly systems based on the use of nitrogen fixing trees such as *Leucaena leucocephala* has potential to bring about favourable changes in soil properties through addition of nutrient rich leaf litter and promoting soil conservation. Practice of growing agricultural crops under scattered trees on farmlands is very old and established practice to fulfil the requirements of shade, fodder, fuel, fruit, timber, vegetable and medicines. Trees like *Acacia nilotica, Acacia catechu, Azadirahcta indica, Dalbergia sissoo, Mangifera indica* are very common in farmer's field as scattered plantation.

Prevalent Agroforestry Systems

Trees on bunds are probably most prominent Agroforestry land use system. Plantation of Eucalyptus and Poplar is the most common bund plantation in the wheat and paddy fields of Haryana and Punjab. In Himachal Pradesh scattered trees of fodder, fuel and timber yielding species are nurtured. The common species are *Grewia optiva, Bauhunia variegata, Toona ciliata, Celtis australis, Prunus puddus* etc. Of lately systematised agroforestry systems have been laid out in farmers field with the efforts of Universities of the region. In Panjab and Haryana Popular and Eucalyptus based Agrisilviculture system, Kinnow based agrihorticulture system and more recently *Melia* based agroforestry system are being adopted by the farmers. In Himachal Pradesh Silvipasture (*Morus*+ Napier, *Grewia* + Napier), Agrisilviculture (*Morus* + wheat), Hortisilviculture (Kinnow + *Leucaena leucocephala*), Hedge row (*Leucaena leucocephala* + *Seteria*). Under agrihorticulture system mango, guava, litchi and citrus fruits are grown with maize, wheat/ mustard and vegetable crops.

Role in Soil and Water Conservation

Agroforestry systems are very effective in arresting soil loss and consequently, the nutrient losses. A long term study (1974 to 1992) in Shivalik foothills on Acacia + bhabbar based silvipastoral land use system reported reduction of soil loss up to 0.2 t ha^{-1} and runoff up to 1.8 per cent of rainfall (Grewal *et al.*, 1996). Grewal (1993) has compared the soil conservation potential of different agroforestry land use systems (Table 1). The study revealed that Eucalyptus-Bhabbar grass (*Eulaliopsis binata*) system was the most effective with soil loss being as negligible as 0.07 t ha^{-1} followed by *Acacia catechu*-Napier grass (0.24 t ha^{-1}) and Teak-Leucaena-Bhabbar (0.43 t ha^{-1}).

Table 1: Effect of different Agroforestry land use systems on runoff soil and nutrient loss in Shivalik region.

Land use system	Soil loss		Runoff		Nutrient loss			
					Nitrogen		Potassium	
	t ha^{-1}	% reduction over CT*	%	% reduction over CT	kg ha^{-1}	% reduction over CT	kg ha^{-1}	% reduction over CT
Cultivated fallow (CT)	5.65	0.00	23.0	0.00	51.30	0.00	5.00	0.00
Eucalyptus-Bhabbar	0.07	98.8	0.05	99.8	0.46	99.1	0.90	82.0
Acacia catechu-forage grass	0.24	95.8	2.00	91.3	6.97	86.4	0.52	89.6
Leucaena-Pennisetum purpureum grass	0.28	95.0	4.40	80.9	6.60	87.1	1.20	76.0
Teak-*Leucaena*-Bhabbar	0.43	92.4	3.30	85.7	2.08	96.0	0.55	89.0
Eucalyptus-Leucaena-Curcuma longa	0.59	89.6	2.60	88.7	2.47	95.2	0.73	85.4
Poplar-*Leucaena*-Bhabbar	1.54	72.7	4.80	79.1	5.90	88.5	1.10	78.0

Source: Grewal, 1993 * Control

Water availability is often more important for crop production than nutrient supplies especially in arid and semi arid zone. Young (1991) have reported reduction in evaporation losses up to 20-30 percent through windbreak plantations. Sharma *et al.* (2001b) also reported that boundary plantation of Poplar had favourable effect on the microclimate. It improved the status of soil moisture between 6-9 m distances and increased the water use efficiency.

Aonla (*Emblica officinalis* var. Chakaiya) based hortipastoral land use systems were evaluated for 12 years (1991-2003). The study conducted by Yadav *et al.* (2005) in Shivalik foothills consisted of the following systems.

T_1 Pure *Emblica officinalis* (No grass)
T_2 *Emblica officinalis* + *Chrysopogon fulvus* (Dholu grass)
T_3 *Emblica officinalis* + *Pennisetum purpureum* (Hybrid Napier grass)
T_4 *Emblica officinalis* + *Cajanus cajan* (Perennial arhar)

The effects of grasses under aonla were compared with pure aonla system. The study on soil loss, runoff and nutrient losses for a decade showed that *Emblica officinalis* + *Chrysopogon fulvus* was the best aonla based silvi-pastoral system for soil conservation. It resulted in 65 and 81 per cent reduction in run off and soil loss, respectively (Fig. 1 & 2). The respective values were 43 and 56, and 19 and 25 per cent for aonla + hybrid napier and aonla + arhar systems, respectively. Aonla + *Chrysopogon* also proved to be the most effective system in reducing nutrient losses and it lead to 68, 66 and 68 per cent reduction in nitrogen, phosphorus and potassium losses, respectively over pure aonla plantation. These systems saved rainwater, soil, N, P_2O_5 and K_2O in the ranges of 4.88 to 30.66 cm, 862-2818 kg ha^{-1}, 3.8-11.0 kg ha^{-1}, 0.005-0.024 kg ha^{-1} and 0.57–1.98 kg ha^{-1}, respectively over pure

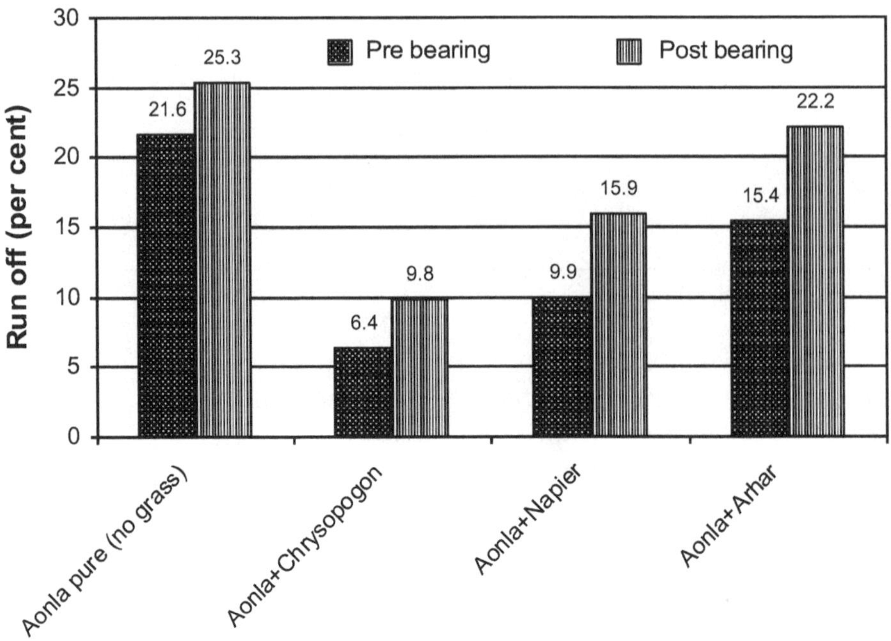

Fig. 1: Run off produced during pre - and post - bearing stages under different silvi pastoral systems (Yadav *et al.*, 2005)

aonla plantation. Effectiveness of aonla + hybrid napier and aonla + arhar in conserving the resources decreased in the post- bearing stage while that of aonla + *Chrysopogon* showed reverse (increasing) trend. The *Emblica officinalis* + *Chrysopogon fulvus* proved to be the most effective cropping system in conserving natural resources on sustainable basis. Solanki and Ram Newaj (2001) also reported that aonla based agroforestry land use system provides more than 60 per cent to ground cover which lowers the soil loss by 7 per cent in Bundelkhand region.

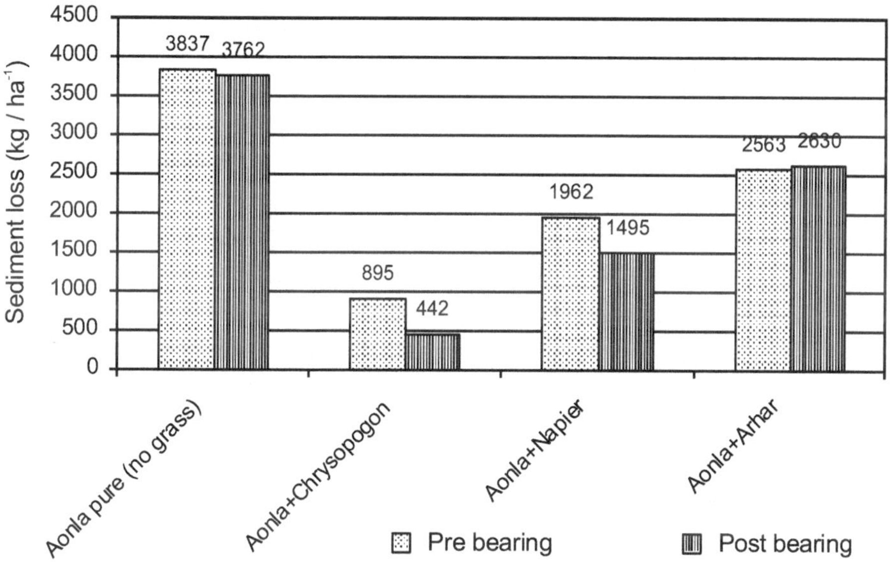

Fig. 2 : Sediment loss during pre- and post-bearing stages under different silvipastoral systems (Yadav *et al.*, 2005).

Role in Soil Amelioration

Agroforestry systems have potential to restore the degraded soils through various ways. It can be due to reduction in run off and soil and nutrient losses, massive biomass production for recycling, nutrient substitution, nutrient mobilization and other direct and indirect effects (Yadav *et al.*, 2003). Ameliorative effects of agroforestry land use were studied in marginal and degraded land in then north-west Uttar Pradesh, now Uttarakhand (Dadhwal and Tomar, 1999). It caused decrease in soil pH by 0.4 units, bulk density by 0.1 mg m^{-3}, improvement in organic carbon content by 0.16 per cent and water holding capacity by 1-5 per cent under the tree canopy in comparison to open area.

Addition of leaf litter in large quantity improved rate of infiltration and permeability under forest land use in comparison to bare land (Sreenivasulu *et al.*, 1996-97).

In Shivaliks, *Leucaena* based land use system have potential to incorporate 183, 10 and 75 kg ha^{-1} yr^{-1} of nitrogen, phosphorus and potassium, respectively through 5 number of lopping every year (Grewal and Dayal, 1993). Its plantation at a spacing of 2m, 3m, 4m and 5m under hedge row system in Agra region added biomass @199, 156, 114 and 92 kg ha^{-1}, respectively (Bhushan *et al.*, 1994). Rao and Ram (1994) recorded potential of nitrogen addition to the tune of 123 kg ha^{-1} when *Leucaena* has planted at a spacing of 4.5m under hedgerow system in degraded black soils of Karnataka. Bhusan *et al.* (1994) and Mittal *et al.* (1992) reported that 50 per cent of nitrogen can be successfully replaced through incorporation of *Leucaena* residues for wheat and maize crops in Yamuna ravines and Shivalik region, respectively. Development of orchard in mid-hill region of Himachal Pradesh reduced nutrient loss and increased infiltration rate in comparison to cultivated land (Sharma *et al.*, 2001a). Addition of organic residues, nutrients and enhanced biological activities upon tree plantation under agro-forestry systems improves soil health through various actions.

Organic carbon

Improvement in organic carbon content consequent to organic residue addition through leaf litter and root decay in large amount is the first step towards soil health improvement. A large portion of added organic residue gets decomposed to release nutrients and some of it gets re-synthesized in to stable organic matter fraction. Effects of various multipurpose trees on soil organic matter content have been compiled in table 2. Nature of tree species, duration of growth, initial soil status and tree growth affect the organic matter accumulation in soil.

Introduction of *Leucaena* in agri-silvi-horticultural system with aonla proved very effective in different soils (Ram Newaj *et al.*, 1999). It increased organic carbon in a range of 28.1 to 62.5 per cent in open canopy and 68.7 to 125.0 per cent below canopy of *Leucaena* (Table 3).

Table 2 : Effect of various Agroforestry land uses on soil organic carbon content

Land use	Years of Experim-entation	Initial OC (%)	Final OC (%)	%age increase in OC	Reference
Leucaena. Leucocephala + Grass	4	0.20	0.81	305.0	Hazra, 1990
Leucaena leucocephala	-	0.60	0.98	63.3	Singh & Gill, 1992
Leucaena leucocephala + Crop		0.48	0.72	50.0	AICRPAF AR, 1999-2000
Prosopis juliflora	7(Site-1) Site-2	0.04 0.02	0.30 0.14	650.0 600.0	Dagar *et al.*, 2001
Prosopis juliflora	20	0.22	0.93	322.7	Yadav *et al.*, 1972
Prosopis cinereria	14	0.22	0.33	50.0	Gupta & Aggarwal, 1992
Prosopis juliflora	20	0.12	0.58	383.3	Singh & Gill, 1992
Acacia nilotica	20	0.12	0.55	358.3	Singh & Gill, 1992
Acacia nilotica	7(Site-1) Site-2	0.04 0.02	0.14 0.08	250.0 200.0	Dagar *et al.*, 2001
Acacia nilotica	-	0.39	0.57	46.2	Hazra, 2000
Acacia nilotica	-	0.60	0.71	18.3	Singh & Gill, 1992
Acacia nilotica	20	0.22	0.85	286.4	Yadav *et al.*, 1972
Albizia lebbek	-	0.39	0.61	56.4	Hazra, 2000
Albizia lebbek	-	0.60	0.68	13.3	Singh & Gill, 1992
Albizia lebbek	20	0.22	0.62	181.8	Yadav *et al.*, 1972
Albizzia lebbeck	20	0.12	0.47	291.7	Singh & Gill, 1992
Albizia lebbek + Crop	-	0.48	0.61	27.1	AICRPAF AR, 1999-2000
Tamarix articulata	7(Site-1) Site-2	0.05 0.03	0.28 0.09	460.0 200.0	Dagar *et al.*, 2001
Eucalyptus	20	0.22	0.66	200.0	Yadav *et al.*, 1972
Eucalyptus terticornis	20	0.12	0.33	175.0	Singh & Gill, 1992
Terminalia arjuna	20	0.22	0.86	290.9	Yadav *et al.*, 1972
Terminalia arjuna	20	0.12	0.58	383.3	Singh & Gill, 1992
Emblica officinalis	-	1.0	1.5	50.0	Solanki & Newaj, 2001

Table 3 : Improvement in organic carbon (%) under aonla and *Leucaena* based agrisilvihorticultural land use system (Initial O.C. 0.32%)

Spacing	Treatment	Black soil O.C. (%)		Mixed red-black soil O.C. (%)	
		Below canopy	Open canopy	Below canopy	Open canopy
10 x 6 m	With *Leucaena*	0.60 (87.5)*	0.41 (28.1)	0.63 (96.9)	0.42 (28.1)
	Without *Leucaena*	0.54 (68.7)	0.42 (31.2)	0.57 (78.1)	0.52 (62.5)
5 x 6 m	With *Leucaena*	0.67 (109.4)	0.44 (37.5)	0.72 (125.0)	0.49 (53.1)
	Without *Leucaena*	0.59 (84.4)	0.45 (40.6)	0.65 (103.1)	0.49 (53.1)

* Figures in parenthesis are percent increase
Source: Ram Newaj *et al.*, 1999

Nutrient content

Researchers have documented the capacity of agroforestry land use systems to improve soil health. A study conducted by Thind and Dhillon (1994) in sandy loam soil of Punjab has recorded significant improvement in available phosphorus and potassium after 3 years of plantation (Table 4). In their study, *Eucalyptus* based land use systems proved better than *Leucaena* based systems in increasing phosphorus and potassium availability. Depth wise studies revealed higher nutrient addition up to 30 cm only.

Table 4 : Effect of Leucaena and Eucalyptus based Land Use System on available phosphorus and potassium in soil profile after 3 years.

Land use systems (LUS)	Available P_2O_5 (kg ha^{-1}) 0-15 cm	15-30 cm	Available K_2O (kg ha^{-1}) 0-15 cm	15-30 cm
Control	15.2	11.5	225	225
Leucaena based	23.2	11.5	300	248
Eucalyptus based	30.5	15.2	413	338
Percent improvement over control under *Leucaena* based	52.6	Nil	33.3	10.2
Percent improvement over control under *Eucalyptus* based	100.7	32.2	83.6	50.2

Source: Thind and Dhillon, 1994

Grewal *et al.* (1996) have recorded nutrient buildup of 173.5, 25.2 and 227.8 kg ha^{-1} of N, P and K, respectively through *Acacia* - Bhabbar based silvi-pastoral land use system after 18 year experimentation during year 1974-1992. Dadhwal *et al.* (1994) have reported that 14 years of popular plantation in loam soils of Dehradun have potential to add 3.8 t ha^{-1} yr^{-1}of litter (leaves and twigs) and supplement 29.4, 29.0, 23.8. 21.3 and 2.8 kg ha^{-1} of N, Mg, Na, K and P, respectively.

An experiment was conducted to evaluate effect of introduction of tree species on fertility status of soil by AICRPAF (Annual Report, 1999-2000) reported that introduction of forestry tree species improved the macronutrient levels in the soil profile in comparison to sole crop and forestry trees alone (Table 5). Plantation of tree species alone failed to improve soil fertility but their combination with crops caused marked increase. Organic carbon content increased from 0.48 under pure crop to 0.61 and 0.72 par cent under shisham and subabul with maize, respectively.

Table 5 : Effect of forestry trees on fertility status of the soil (kg ha^{-1})

Treatment	Available N	Available P	Available K
Sole crop	252.5	20.2	280.5
Subabul alone	180.2 (-28.6)*	12.5 (-38.1)	250.0 (-10.9)
Shisham alone	192.5 (-23.8)	2.5 (-87.6)	260.7 (-7.1)
Subabul + Crop	265.3 (5.1)	23.5 (16.3)	260.7 (-7.1)
Shisham + Crop	283.5 (12.3)	25.5 (26.2)	320.5 (14.3)

* Figures in parenthesis are percent change over sole crop.
Source: AICRPAF Annual Report, 1999-2000

Effect on Soil Physical Properties

Agroforestry land use systems add organic residues to the soil profile through leaf litter which have potential to decrease bulk density and increase in total porosity and infiltration rate (Painuli and Yadav, 1998). However, initially the effect is limited to the surface soil layer. Jain and Garg (1996) also stated that bulk density in 0-15 cm soil layer decreased and water holding capacity increased in the sites under 10 year old plantations as compared to the unplanted site.

Balamurugan *et al.* (2000) studied the influence of *Eucalyptus citridora* plantation on the physical properties of the soil at different depths (0-30, 30-60 and 60-90cm) and reported that plantation have potential to increase clay and silt content and decrease sand content in soil profile. There was decrease in bulk and particle density but increase in porosity and water holding capacity. The change in soil properties was more prominent in 0-30 cm of soil profile than lower depths.

The effects on physical properties are visible over a longer period. *Leucaena* plantation for 21 years showed marked effect on bulk density, porosity, infiltration rate and penetrometer resistance. *Leucaena* plantation for 21 years decreased bulk density by 104 kg m^{-3} and penetrometer resistance by 0.81 but increased porosity by 5.4% and infiltration rate by 1.06 cm min^{-1}. Changes in soil texture were very small but trend was towards higher content of finer fractions.

Effect on Soil Chemical Properties

Beneficial effects of tree plantation on alkali soils were due to sharp reduction in ESP and soil pH where efficiency of diiferent tree species was in the order of *Tamarix articulate* > *Prosopis juliflora* > *Acacia nilotica* (Dagar *et al.*, 2001; Table 6). Results from other studies are similar in nature and are summarised in the table.

Effect on Soil Microbial Properties

The most favourable advantage of agroforestry land use systems is increase in organic matter of the soil profile through addition of leaf litter which favours growth and multiplication of various agriculturally important micro organisms like bacteria, fungi, actinomycetes etc. These soil microbes play very important role in mineralization and availability of various plant nutrients in soil profile. Soil micro organisms also have potential to produce growth promoting substances. More litter accumulation under high humidity and moisture conditions existing below the tree canopy favours the growth and multiplication of microbial population in the soil profile below tree canopy. N-fixing trees species like *Leucaena* and *Acacia* have better potential of improvement in the microbial biomass through symbiotic association of Rhizobium bacteria.

Scope of Agroforestry Land Use Systems

Increasing population of the country squeezed per capita availability of natural resources. Crop diversity through agro-forestry systems has ample potential to meet the requirements of food, fodder and fuel for the densely populated countries like India. Agroforestry tree has great potential to rehabilitate degraded and problematic soil through buffering effect of soil pH towards neutral and improvement in fertility status of soil, which further spread its adoptability for the barren and underproductive land.

Table 6 : Effect of forest tree plantation on physical properties of alkali soil of semiarid region

Land use	Year of exp-eriment ation	pH			EC (dSm^{-1})			CEC(%)			Reference
		Initial	Final	Change	Initial	Final	Change	Initial	Final	Change	
A. nilotica	-	7.8	7.4	-0.4	0.41	0.30	-0.11				Hazra, 2000
A. nilotica	-	7.7	7.5	-0.2	0.28	0.22	-0.06	14.3	15.2	0.9	Singh & Gill, 1992
A. nilotica	20	10.2	8.4	-1.8	1.18	0.25	-0.93				Yadav *et al.*, 1972
A. nilotica	20	10.3	9.0	-1.3							Singh & Gill, 1992
A. nilotica Site 1& 2	7	10.5 10.6	10.1 10.3	-0.4 -.03	3.20 1.35	0.88 1.23	-2.32 -0.12				Dagar *et al.*, 2001
A. lebbeck	-	7.8	7.2	-0.6	0.41	0.27	-0.14				Hazra, 2000
A lebbeck	-	7.7	7.6	-0.1	0.28	0.28	0.0	14.3	14.8	0.5	Singh & Gill, 1992
A. lebbeck	20	10.2	7.9	-2.3	1.18	0.32	-0.86				Yadav *et al.*, 1972
A. lebbeck	20	10.3	8.67	-1.63							Singh & Gill, 1992
Leucaena+ Grass	4	7.2	7.0	-.2	0.27	0.11	-0.16				Hazra, 1990
Leucaena	-	7.7	7.2	-.5	0.28	0.18	-0.10	14.3	15.6	1.3	Singh & Gill, 1992
P. juliflora	20	10.2	7.5	-2.7	1.18	0.51	-0.67				Yadav *et al.*, 1972
P. juliflora	20	10.3	8.0	-2.3							
P. juliflora *Site1 & Site 2*	7	10.4 10.5	9.7 10.1	-0.7 -0.4	3.15 1.20	0.67 1.02					Dagar *et al.*, 2001
Eucalyptus	20	10.2	8.5	-1.7	1.18	0.44	-0.75				Yadav *et al.*, 1972
Eucalyptus terticornis	20	10.3	9.18	-1.12							Singh & Gill, 1992
T. arjuna	20	10.2	7.9	-2.3	1.18	0.32	-0.86				Yadav *et al.*, 1972
T. arjuna	20	10.3	8.15	-2.15							
Tamarix articulata *Site 1 &2*	7	10.4 10.5	9.2 9.9	-1.2 -0.6	3.05 1.13	0.36 0.50	-2.69 -0.63				Dagar *et al.*, 2001

Source: Dagar *et al.*, 2001

References

Aggarwal, R.K., R.P.Yadav and Pawan Sharma (2000). Integrated nutrient supply system for sustainable production in rainfed shivalik. In: *Fifty Years of Research on sustainable Resource Management in Shivaliks* (eds. S.P.Mittal, R.K.Aggarwal and J.S.Samra, CSWCRTI, R.C. Chandigarh). pp. 377-392.

AICRPAF (AR). (1999-2000). All India Coordinated Research Project on Agroforestry. pp. 45.

Balamurugan, J., K. Kumaraswamy and A.Rajarayan (2000). Effect of *Eucalyptus citridora* on the physical and chemical properties of soil. *J. of Indian Soc. Soil Science* 48(3):491-495.

Bhushan, L.S., Om Prakash, S. Singh and J.S. Deshwal (1994). Agro forestry system for sustainable production in Yamuna ravines. *In:* Agroforestry: Traditions & Innovations. (eds Pratap Narain, Dadhwal, K.S. and Singh, R.K.), CSWCRTI, Dehradun, pp. 33-39.

Dadhwal, K.S., A.Raizada, Pratap Narain and D.P. Datta (1994). Agro forestry systems of Doon valley and plains of western Uttar Pradesh. *In:* Agroforestry: traditions & Innicvations (ed. Pratap Narain, Dadhwal, K.S. and Singh, R.K.), CSWCRTI, Dehradun, pp. 12-18.

Dadhwal, K.S. and V.P.S. Tomar (1999). Enhancing production by managing tree roots in an agroforestry land use system. *SAIC Newsletter,* Dhaka, Bangladesh, 9:5 & 7

Dagar, J.C., G. Singh and N.T. Singh (2001). Evaluation of forest and fruit trees used for rehabilitation of semiarid alkali soil in India. Arid Soils and Rehabilitation.

Gupta, J.P. and R.K. Aggarwal (1992). Integrated resource management for sustained crop production in arid regions. *American J. Alternative Agriculture* 7:157-161.

Grewal, S.S. (1993). Agroforestry system for soil and water conservation in Shivalik. *In:* Agroforestry in 2000 A.D. for semi arid and arid tropics, N.R.C.A.F., Jhansi, India. pp. 82-85.

Grewal, S.S. and S.K.N. Dyal (1993). *Leucaena leucocephala* hedge row as a source of organic manure, *Ind. J. Soil Cons.* 21(2):69-71.

Grewal, S.S., Kehar Singh, M.L.Juneja and Shiv Charan Singh (1996). Relative growth, fuel-wood yield, litter accumulation and conservation potential of seven acacia spp. on a sloping bouldry soil. *Indian J. Forestry* 19: 174-182.

Hazra, C.R. (1990). Forage from agroforestry based based production systems. *In:* Proc. Int. Sym. Water Erosion, Sedimentation and Resource Conservation, Dehradun, India, pp. 436-446.

Hazra, C.R. (2000). Woody perennial in farming systems for enhanced production and livelihood. *In:* Advances in Land Resource Management for 21st Century, Soil Conservation Society of India, pp. 341-358.

Jain, R.K. and V.K.Garg (1996). Effect of decade old tree stands on some properties of soils while re-vegetating sodic wastelands. *Indian Forester*.122: 464.

Mittal, S.P., S.S.Grewal, Y.Agnihotri and A.D. Sud (1992). Substitution of nitrogen requirements of maize through leaf biomass of *Leucaena leucocephala*-Agronomic and Economic consideration. *Agroforestry Systems* 19:207-216.

Painuli, D.K. and R.P.Yadav (1998). Tillage requirement of Indian Soil. *In:* 50 Years of Natural Resource Management Research (eds G.B. Singh and B.R. Sharma). Division of Natural Resource Management, ICAR, New Delhi, pp. 245-262.

Ram, Newaj, S.K.Shukla and R.S.Yadav (1999). Varietal evaluation of aonla (*Emblica officinalis*) in agri-silvi-horticultural system under rainfed conditions on marginal lands. Annual Report-1998-99, NRCAF, Jhansi, pp. 44-46.

Rao, M.S. and N.S.Jaya Ram (1994). Tree based farming systems: An appraisal of research works of Bellary region. *In:* Agroforestry: Tradition & Innovations. (eds. Pratap Narain, Dadhwal, K.S. and Singh, R.K.), CSWCRTI, Dehradun, pp. 76-84.

Sharma, I.P., Sushil Kumar and J.C. Sharma (2001a). Rainfall-runoff-soil loss relationships under different land use in mid hill region of Himachal Pradesh. *Indian J. Soil Cons* 28(2):91-97.

Sharma, N.K., J.S.Samra and H.P. Singh (2001b). Poplar (*Populus deltoids*) based agroforestry systems for an alluvial soil under irrigated condition in western Uttar Pradesh. *Indian Forester*. January, 2001. pp. 61-69.

Singh, G. and H.S. Gill (1992). Ameliorative effect of tree species on characteristics of sodic soils at Karnal. *Indian J. Agric. Sci.* 62: 144-146.

Singh, H.P., K.L.Sharma, B. Venkateshwarlu, T.Vishnumurthy, and K. Neelaveni (1998). Prospectus of Indian agriculture with special reference to nutrient management under rainfedcondition. Proc. National Workshop on Long Term Soil Fertility management through Integrated Plant Nutrient Supply, IISS, Bhopal, India, pp.34-53.

Solanki, K.R. and Ram Newaj. (2001). Rehabilitation of degraded lands through agroforestry. *Indian Farming* 50(12):31-35.

Sreenivasulu, V. Omkar, S.P. Rai and D.K. Aggarwal (1996-97). Soil physico-chemical porpeeties of Basantar catchments, J&K. Report No. CS-28. National Institute of Hydrology, Roorkee.

Thind, H.S. and M.S. Dhillon (1994). Comparative effect of Leucaena and Eucalyptus on soil fertility and productivity of light textured wastelands. *In: National conference on soil and water conservation for sustaining production and panchayat raj.* held at Chandigarh, India from January 28-30, 1994.

Yadav, J.S.P., D.R. Bhumbla and O.P. Sharma, (1972). Performance of certain forest species on a saline alkali soil. *In:* Proc. Int. Symp. On 'New developments in the Field of Salt Affected Soils. Cairo, pp. 683-690.

Yadav, R.P., R.K. Aggarwal and S.P.Mittal (2003). Restoration of productivity of degraded lands through alternative land uses and recycling of organic residues. *In:* Alternative Land Use Systems for Sustainable Production (eds. Prof. S.P.Mittal), published by Oriental Enterprises, Dehradun. pp. 435-454.

Yadav, R.P., R. Prasad, R.K. Aggarwal and J.S. Samra (2005). Aonla (*Emblica officinalis*) based silvi-pastoral systems for soil and water conservation in degraded lands of Shivalik foothills. *Indian J. Soil Cons.* 33(3): 207-212.

Young, A. (1991). Agroforestry for soil conservation. CAB International (ICRAF). pp.9.

Chapter-14

Land Degradation in the Shivaliks of Jammu and Kashmir : Role of Agroforestry in Conservation and Reclamation

VIKAS SHARMA[1] AND S. K. GUPTA[2]
[1]Division of Soil Science and Agricultural Chemistry and [2]Division of Agroforestry, Faculty of Agriculture, Sher-e-Kashmir University of Agricultural Sciences and Technology-Jammu, Chatha – 180009.

ABSTRACT

The lower Shivaliks of Jammu and Kashmir comprise 12% of the total area of Jammu region. The productivity of the soils is not only low but highly unstable. High intensity monsoonal rains coupled with rolling topography, low organic matter and coarse nature of the soils are primarily responsible for soil erosion in the region. Soil loss due to erosion depletes the soil nutrients and, is primarily responsible for land degradation, low productivity and poor economic status of the farmers in the region. Further, the adoption level of the farmers regarding modern conservation techniques is also low. Though soil erosion can be controlled or minimized through agronomical and/or mechanical techniques but agroforestry intervention is one such option which can be used for managing the land in the problematic conditions. Agroforestry systems have proven potential of conserving soil and water besides enrichment of soil through biomass additions. Agroforestry systems, therefore, help curtail soil erosion both by intercepting the rain and minimising its erosivity as well as by reducing the erodibility of the soil. Tree based cropping systems enhance the organic content in soils and therefore, create more stable aggregates, which provide more resistance to erosive forces. The chapter discusses the importance of agroforestry in the fragile Shivaliks of Jammu and Kashmir and suggest how to conserve and restore degraded lands in the region.

Introduction

The Shivaliks covering an area of 2.14 m ha fall in four north-western states of India i.e. Punjab (0.14 m ha), Haryana (0.06 m ha), Himachal Pradesh (1.14 m ha) and Jammu and Kashmir (0.80 m ha) (Singh, 2010). The lower Shivaliks or the foothill Himalayas fall under the Jammu division of Jammu and Kashmir state, extending from district Kathua in the south-east to Rajouri in the north-west, constituting about 12% of the total area of Jammu region (Fig. 1). It represents the most fragile ecosystem of Himalayan mountain range because of its peculiar geological formations and highly erodible soils (Singh and Khera, 2009). The productivity of the soils of the lower Shivalik region of Jammu is not only low but highly unstable. Soil erosion due to huge runoff results in considerable soil and nutrient loss and, is primarily responsible for land degradation, low productivity and poor economic status of the farmers in these rainfed areas. The loss of top soil result in vanishing of nutrients essential for plant growth, which influence the crop productivity. Soil erosion in the Shivalik or Southern Himalayas is aggravated by massive deforestation and severe denudation of the hills and causing severe floods in the plains besides removal of more than 80 tonnes of soil ha^{-1} $year^{-1}$ (Sharma, 2004). In some cases the soil loss in Shivaliks due to erosion has been reported up to the tune of 400 t ha^{-1} yr^{-1} (Sidhu *et al.*, 2007). Loss of plant nutrients, however, is accelerated due to faulty land use pattern. The top soils from forest and well managed cultivated land contained 0.83 and 0.58 per cent organic carbon, respectively, as compared to 0.40 and 0.30 per cent from the barren and unmanaged cultivated lands (Gupta, 1994). Due to fragility, marginality, low accessibility and resource heterogeneity, there is little scope for enhancing crop productivity with the existing infrastructural facilities and the level of adoption of technologies.

Physiography and Climate

The lower Shivaliks of Jammu and Kashmir are classified under the physiographic unit Hills ranging from 320 to 720 meters above mean sea level (Rashid and Arora, 2007). The area is mostly dominated by Shivalik, Muree and Subathu group of sediments. The general

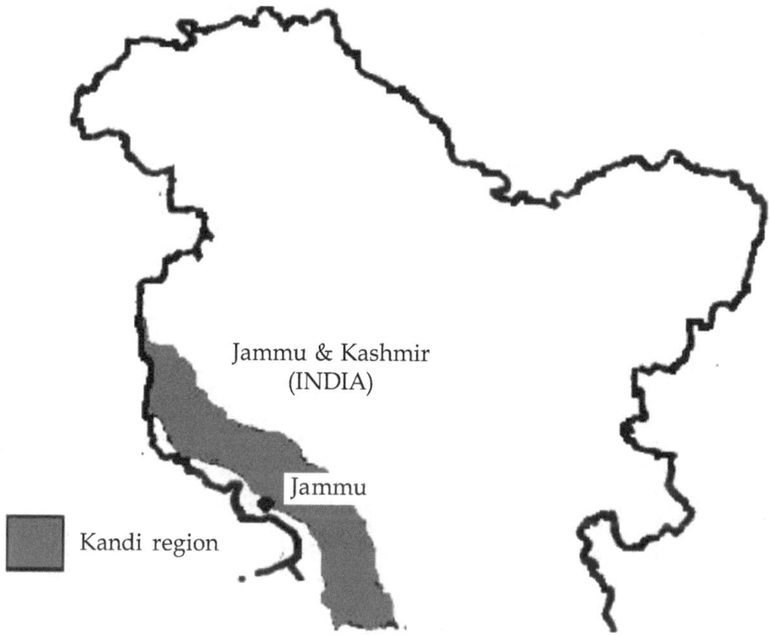

Fig. 1 : *Kandi* belt of Jammu and Kashmir

lithology is sandstone, conglomerate, shale, silt stone and limestone. The rainfed area of Jammu Shivaliks has unique land, soil and climatic features representing a semi-arid or sub-humid type of climate. However, the climate of the whole area is characterized by sub-tropical nature. The region is characterized by erratic rainfall (750-1200 mm annually), undulating terrain, frequent droughts, low soil organic matter and coarse textured soils. The soil surface is infested with sand stones and extremely poor water retention capacity. The agricultural activity in the region is totally dependent on monsoonal rains. About 80 % of the rains occur in first two months of monsoon season (July-August), with high intensity rainstorms followed by staggered rains (Fig 2). The physiographic characteristics of the area is itself a major factor contributing to the continuous degradation of these catchments. The weak lithology of the lower Shivaliks consisting of rocks like sandstone, conglomerate, shale, silt stone and limestone are relatively easily weatherable and therefore prone to quick erosion. The topography of the region ranging from gently sloping to moderately-steep sloping retards the vertical development of soils. Due to great variation in steepness of the slope in this region, erosion is the major problem of land husbandry (Gupta and Gupta, 2004). Sloping relief pattern results in accelerated erosion causing removal of surface material resulting in lesser percolation of water which is essential for profile development. This also causes less availability of water for the growth of plants and woody vegetation which help in checking soil erosion. The average temperature during summer, winter and rainy seasons varies between 18°C and 40°C, 4°C and 23°C and 14°C and 32°C, respectively.

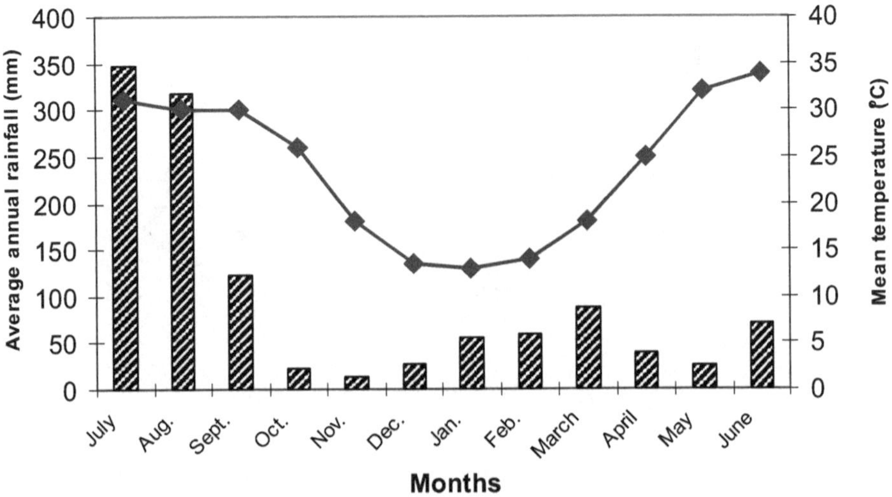

Fig. 2 : Monthly average rainfall and temperature in Jammu region (mean of 20 years)

Soil Loss and Land Degradation

The entire lower shivaliks are drained by a number of major and minor rivers including Chenab, Tawi, Basantar and Ujh. Apart from these, there are ephemeral streams (*khads*) like Devak, Bain, Bhagar and Sahar dotting the landscape. During the rainy season the soil erosion is heavy, resulting in heavy loss of nutrients and top fertile layer. Comparison of crop yields have shown that average productivity of various crops is poor in *kandi* region compared to the state or national average (Table 1). The degradation of catchments in terms of land, soil and vegetation is enormous and is the result of both natural and anthropogenic factors.

Natural Factors

This high intensity rainfall is capable of disturbing the top soil, and in turn loosening it, resulting in sheet and rill erosion. The problem is further aggravated as the soil in the area is coarse textured and prone to erosion (Sharma *et al.*, 2009). It is generally considered that as the silt (0.002-0.05 mm) or silt and very fine sand (0.05-0.10 mm) fraction increases, soil erodibility increases (Wischmeier and Mannering, 1969) because of binding affect of clay, transportability of fine and non-aggregate particles and the detachability of sand and silt (Le-Bissonnais, 1996).

Table 1 : Comparison of major crop yields of *kandi* region of Jammu

Crop	Yield (kg ha^{-1})		
	Lower Shivaliks (*Kandi* region)	J&K state	All India
Wheat	890	1530	2583
Maize	1205	1710	1755
Coarse cereals	1165	1617	1065
Pulses	450	564	622

Source : Sharma and Sharma, 2003

The extent of rill soil erosion can be discerned in number of broad longitudinal valleys known as *Duns* like Udhampur, Jhajjar Kotli and Dansal in lower shivaliks of Jammu region. The seasonal high intensity rains also results in sudden and flash floods in rivers, *nallahs* and *khads* fed by their respective catchments. The water loaded with sediments from the catchments exerts enormous pressure on the banks resulting in stream bank erosion resulting in loss of sizable chunks of productive land over time.

Anthropogenic Factors

Increased pressure of human and cattle population and decreased size of land holdings in the area have resulted in indiscriminate felling of trees, removal of bushes and grazing and browsing. The human population of the state increased at an annual compound growth rate of 2.64% as against 1.08% ACGR of food grains between 1961 and 2001 (Table 2) (Sharma and Sharma, 2003). This indicated that while increase in the population was substantial, the food grain production did not cope with this increase. The situation is still worse in the *kandi* region which faces acute shortage of food grains and forage resources.

Table 2 : Population and food grain production in J&K state

Year	Population	Food grain production (million tonnes)
1961	3.56	0.96
1971	4.61	1.00
1981	5.95	1.17
1991	7.72	1.32
2001	10.12	1.48

Source : Sharma and Sharma, 2003

Lack of knowledge regarding soil and water conservation measures has also contributed towards land degradation in the area. It has been observed that only about 43 per cent of the farmers in the lower Shivaliks of Jammu are aware about the severity of the problems related to runoff and soil loss (Sharma *et al.*, 2004). Only 31 per cent of the farmers adopt any of the soil and moisture/water conservation measures in the area. More than 76 per cent of the farmers are cultivating land along the slope resulting in soil, nutrient and water loss as only 24 per cent of the respondents plough across the slope.

The community in these areas faces a lot of constraints in adoption of improved soil and water conservation measures. About 76.6 per cent of the farmers feel that they had lack of skill in handling the practices like contour bunding, staggered trenches etc. (Table 3) (Sharma *et al.*, 2005). About 95 per cent of the inhabitants had problems due to undulating topography which causes problem in application of conservation practices. They had to rely more on human labour rather than machinery because of sloping and fragmented fields. Further, poor socio-economic condition of the farmers restricts them from employing any labour or using any machinery.

The land holdings in the *kandi* region are small and fragmented. A sample survey in the area showed that 92% of the families have cultivable holdings of less than 2 ha and 80% have less than 1.5 ha (Arora *et al.*, 2006). The tenancy law of transferring the land to all children of the family has made the matter still complicated. At present, the land holding exists in small to large categories of < 1.0 ha, 1 to 5 ha and > 5 ha in the region. Though the family size is bigger in the region, most of the members of the family go for job or labour work to earn their bread instead of performing agriculture practices on their small piece of lands.

Table 3 : Constraints in adoption of soil and rainwater management practices (n=120)

Constraints for adoption of soil and rain water conservation practices	No. of Farmers	Per cent
Non-availability of design and implements used for rainwater harvesting	107	89.20
Lack of technical knowledge	92	76.60
Topographical problems	114	95.00
Ignorance of rainwater management practices	69	57.50
Small & fragmented land holdings	102	85.00
Poor socio-economic condition	105	87.50

Source : Sharma *et al.*, 2005

In order to meet the demand of food, fiber, fuel and fodder, due to exponential increase in population pressure, there has been large scale deforestation. It has led to unabated soil loss and land degradation. Forest cover plays significant role not only in checking the soil erosion and run-off losses but also maintains the balance in physical landscape through the reclamation of degraded and eroded lands.

Agroforestry in Soil Conservation

Soil erosion can be controlled or minimized through agronomical and/or mechanical measures. Agroforestry is one such system which can be used for managing the land in erosion prone areas. Agroforestry is a special case of mixed-cropping, whereby food crops and animals are raised in association with trees, shrubs and other woody perennials (Lal, 1990). Agroforestry systems take advantage of trees for many uses, to hold the soil, to increase fertility through nitrogen fixation, or through bringing minerals from deep in the soil and depositing them at the surface by leaf-fall (Martin and Sherman, 1992).

Runoff and soil loss from Agroforestry Systems

Established hedgerows of *Leucaena* at 2-m spacing were very effective in reducing water runoff and controlling erosion (Lal, 1989).

They observed a mean seasonal erosion of 4.3, 0.57, 0.10, and 0.60 t/ha from maize for plow-till, *Leucaena*-4m, *Leucaena*-2m, *Gliricidia*-2m treatments, respectively. Mean runoff in the first season from treatments listed in the order above was 17.0, 4.9, 3.3, and 2.4 percent of the rainfall received.

Alley cropping with *Leucaena, Gliricidia* and grass barriers has been found effective for mulching and erosion control on sloping lands up to 30 percent. During a nine year study period (Narain *et al.*, 1997), the average annual monsoon rainfall of about 1000 mm, in western Himalayas caused 347 mm runoff and 39 Mg ha^{-1} soil loss per year from fallow land. Contour tree-rows or leucaena hedges reduced the runoff and soil loss by 40% and 48%, respectively, over the maize plot, reducing soil loss to about 12.5 Mg ha^{-1}. This reduction in erosion was primarily due to the barrier effect of tree or hedgerows and micro-terraces formed through sediment deposition along the contour barriers. The total sediment deposition along the hedge and tree rows increased considerably with consequent reduction of soil loss (Table 4).

Table 4 : Sediment deposition along vegetative barriers and soil loss from different landuse systems

Treatment	Average soil loss (t $ha^{-1}yr^{-1}$)	Soil deposition (t $ha^{-1}yr^{-1}$)
Leucaena hedges in turmeric field	15.77	5.57
Leucaena hedges in maize field	51.33	12.09
Leucaena trees in maize field	28.50	8.82
Leucaena trees in turmeric field	10.01	6.78
Eucalyptus trees in maize field	20.62	5.80
Eucalyptus trees in turmeric field	11.52	7.12
Eucalyptus + maize - wheat	-	7.30
Leucaena + maize - wheat	-	11.23
Cultivated fallow	-	39.01

Source : Narain *et al.*, 1997

In the Shivalik ecosystem, agroforestry systems has proven to be more conservation effective than traditional crops on marginal soil Season runoff with 816 mm rainfall in the foot hill Himalayas was 4.4,

11.2, 20.5 and 23.0% whereas soil erosion loss 0.28, 1.03, 2.69 and 5.63 tonnes, N loss 6.6, 19.5, 42.5 and 51.3 kg and K loss 1.2, 1.8, 3.0 and 5.0 kg $\cdot ha^{-1} \cdot yr^{-1}$, from *Leucaena leucocephala* (K-8) for fuelwood at 2 m × 2 m spacing in upper and napier grass (*Pennisetum purpureum*) for fodder in lower canopy, *Leucaena leucocephala* (K-8) for fuelwood at 2 m × 2 m spacing in upper and Leucaena (El Salvador) at 0.5 m × 0.5 m spacing for fodder in lower canopy, rainfed crop sequence of sesame followed by rapeseed and cultivated fallow, respectively (Grewal *et al.*, 1994).

Soil erodibility

Soil erodibility is the susceptibility of the soil to erosion. Although a soil's resistance to erosion depends in part on topographic position, slope steepness and the amount of disturbance, still the soil properties play an important role in governing the resistant offered to erosive forces (Morgan, 1995). The soil erodibility depends primarily on the physical characteristics of the soils viz., nature and amount of soil aggregates, organic matter content and particle size distribution (Singh and Khera, 2008).

A study in the Punjab Shivaliks has revealed that in the north-eastern part, slope steepness, slope length, convexo-concave and concave slope patterns and sparse vegetation were the dominant factors governing erosion hazard whereas, in the central and lower south-western part of the area, high soil erodibility and sparse vegetation were the most important factors (Kukal *et al.*, 1991). They observed that the erodibility of the soil was strongly correlated (r= 0.98) with amounts of silt + very fine sand.

Land use pattern has a significant impact on these physical parameters of soil. Erodibility of the soils of foot hill Shivaliks of Jammu could be arranged in the order: barren uncultivated lands> cultivated un-managed lands> cultivated well managed lands> forest lands (Gupta *et al.*, 2005). Thus, soils of forest lands appeared less susceptible to erosion than soils of other groups. They recorded lower values of suspension and percolation ratio in forest soils than those of others (Table 5). Greater these and values, the more readily easily the soil could be eroded (Datta *et al.*, 1990). Lessons can be leant from the natural systems and by introducing tree based agricultural systems further degradation of this fragile eco-system can be prevented.

Table 5 : Erodibility indices of soils under different land use type.

Parameter	Barren Uncultivated Lands		Cultivated Unmanaged Lands		Forest Lands	
	Surface	Sub - Surface	Surface	Sub - Surface	Surface	Sub - Surface
Erosion Ratio	127.97	148.74	63.72	82.95	24.18	32.64
Dispersion Ratio	52.10	60.72	38.36	46.37	21.17	26.16
Clay Moisture Equivalent Ratio	0.52	0.48	0.61	0.58	0.90	0.85
Erosion Index	105..30	126.48	50.05	69.03	21.70	27.59
Clay Ratio	14.57	17.10	6.63	8.08	3.95	4.48
Suspension Percentage	15.85	16.55	14.61	15.18	9.65	11.50
Suspension Ratio	210.62	249.05	99.77	137.41	43.41	55.34
Percolation Ratio	38.27	37.03	24.21	21.68	10.74	15.10

Source : Gupta *et al.*, 2005

The dispersion ratio reduced, whereas, the water stable aggregates (>0.5 mm) increased under a poplar based Agroforestry system in the foothill Shivaliks irrespective of the soil type as compared to a non-Agroforestry system of cultivation (Table 6) (Gupta *et al.*, 2006). Stable the soil aggregates greater will be the resistance offered by them to erosive forces. Clay content, water stable aggregates (WSA), mean weight diameter (MWD), geometric mean weight diameter (GMD) and organic carbon (OC) were negatively correlated with soil loss and soil erodibility (K) values (Singh and Khera, 2007).

Carbon sequestration/ Biomass additions

Agroforestry is generally practiced for checking soil erosion, land reclamation and increased productivity. At the same time it adds sufficient biomass into the soil, which has been shown to improve the physical as well as chemical properties thereby reducing the erodibility of soil (Gupta *et al.*, 2006). Agroforestry systems are believed to increase or at least maintain the organic matter level of soils (Young, 1989) mainly through litterfall.

Table 6 : Effect of Agroforestry system on soil erodibility

Land use	Surface soils				Sub -surface soil			
	Loamy sand	Loamy	Sandy clay loam	Mean	Loamy sand	Loamy	Sandy clay loam	Mean
	Dispersion ratio							
Agroforestry	3.4	2.4	2.1	2.6	11.0	4.5	4.7	6.7
Non-Agroforestry	15.4	8.3	6.3	10.0	14.8	8.2	6.9	10.0
	Water stable aggregates >0.5 mm							
Agroforestry	10.6	2.0	11.1	7.9	8.0	5.0	7.0	6.7
Non-Agroforestry	2.0	0.51	5.6	2.7	0.57	0.16	2.0	0.92

Source : Gupta *et al.*, 2006

Agroforestry interventions improve aggregation of soil through huge amounts of organic matter in the form of leaf biomass. Soils under forest land use are more stable as compared to the soils under cultivation due to the higher organic carbon content resulting in more stability of soil aggregates (Kukal *et al.*, 1993). The carbon harvesting potential of the trees and the biomass input (in terms of litter fall) to the soil thus is an important factor reducing erodibility of soils, apart from sequestering carbon. The extent of improvement may be affected by the age of the trees and the soil type. The soils under agroforestry had 2.9–4.8 Mg ha^{-1} higher soil organic carbon than in sole crop (Table 7) (Gupta *et al.*, 2009). The impact of increased carbon was visible on the mean weight diameter (MWD) and water stable aggregates (WSA) of soil, which increased by 3.2, 7.3 and 13.3, and 14.4, 32.6 and 56.9 times in soils with 1, 3 and 6 years plantation, respectively in comparison to sole crop. The magnitude of associated organic carbon decreased with aggregate size (Saha *et al.*, 2010). Climate change has posed a question whether the planet is really warming, and how to mitigate and adapt to its impacts. To overcome the problems of possible climate changes, the Prime Minister released India's National Action Plan on Climate Change in New Delhi in 2008 wherein it has been emphasized that climate change is a global challenge and can only be successfully overcome through a global, collaborative and cooperative effort. The conservation of soil and water, sustaining Himalayan ecosystem, sustainable agriculture need to be focused attention.

Table 7 : Influence of poplar based Agroforestry on total soil organic carbon (Mg ha^{-1})

Landuse	Years		
	One	Three	Six
	0-15 cm		
Agroforestry	12.4	14.6	15.8
Sole crop	9.0	9.7	9.2
	15-30 cm		
Agroforestry	10.8	11.3	13.3
Sole crop	8.1	8.3	8.1

Source : Gupta *et al.*, 2009

Removing atmospheric carbon (C) and storing it in the terrestrial biosphere is one of the options, which have been proposed to compensate greenhouse gas emissions (Albrecht and Kandji, 2003) and minimize the risks of climate change. Soil is major sink for the C, and as such introduction of tree based cropping systems can enhance the rate of C sequestration in the soils. Agroforestry system thus can play a major role in minimizing environmental degradation, both in terms of land and atmosphere.

Future Challenges

Keeping in view the poor socio-economic status of the farmers and the inherent causes of soil erosion and land degradation prevalent in the area, the challenge lies in protecting the limited land resource from further degradation, improving their productivity and checking poverty in *kandi* region of the state. To rehabilitate the area, it is required to conserve soil against erosion, manage rainwater efficiently to produce more crops, to provide food, fodder and fuel. Traditionally practiced Agroforestry needs to be developed on scientific lines to produce more biomass per unit area without further degradation of resources. Alternate land use systems of alley cropping, agrisilviculture, silvipastoral and agrihorticulture are recommended in the foot hills of

Shivaliks of Jammu region, hilly region to optimize the use of natural resources, minimize the use of inputs derived from non-renewable resources and reduce the risk of environmental degradation (Sharda, 2005). Agroforestry has gained popularity among farmers, researchers, policy makers and other for its ability to contribute significantly in meeting deficit of tree products, socio-economic and environmental benefits (Dhyani *et al.*, 2009). Horticulture has a good scope in the area. The possibility of combining horticulture and forestry with agricultural crops, grasses and fodders needs to be further explored on scientific basis. Agroforestry is an important land use option for resource poor region, that can not only minimize the rate of land degradation but can also be used to reclaim erdoded/marginal lands while at the same time increase livelihood security and reduce vulnerability to climate and environmental vagaries.

References

Albrecht, A., and S.T. Kandji (2003). Carbon sequestration in tropical agroforestry systems. *Agriculture, Ecosystems & Environment*, 99(1-3), 15–27.

Arora, S., V. Sharma, A. Kohli and V.K. Jalali, (2006). Soil and water conservation for sustainable crop production in Kandi region of Jammu. *Journal of Soil and Water Conservation*, 5(2), 77-82.

Datta, M., P.K. Saha, and H.P. Chaudhury (1990). Erodibility characteristics of soils in relation to soil characteristics and topography. *Journal of the Indian Society of Soil Science*, 38(3), 495-498.

Dhyani, S.K., R. Newaj, and A.R. Sharma, (2009). Agroforestry: its relation with agronomy, challenges and opportunities. *Indian Journal of Agronomy*, 54(3), 249–266.

Grewal, S., M. Juneja, K. Singh and S. Singh (1994). A comparison of two agroforestry systems for soil, water and nutrient conservation on degraded land. *Soil Technology*, 7(2), 145-153.

Gupta, G.D. (1994). Characterization of eroded soils of kandi belt of Jammu district, Jammu and Kashmir state (M.Sc. Thesis). Sher-e-Kashmir University of Agricultural Sciences and Technology of Jammu.

Gupta, R.D., and G.D. Gupta (2004, June 4). Soil erosion in J & K. Daily Excelsior. Jammu.

Gupta, G.D., R.D.Gupta, D. Kher and N.M. Sumeria (2005). Studies on erodibility indices in relation to soil properties in kandi belt of Jammu district. *Journal of Research, SKUAST–J*, 4(2), 209-213.

Gupta, N., S.S. Kukal and P. Singh (2006). Soil erodibility in relation to poplar based Agroforestry system in north western India. *International Journal of Agriculture and Biology* (Pakistan), 8(6), 859-861.

Gupta, N., S.S. Kukal, S.S. Bawa and Dhaliwal, G.S. (2009). Soil organic carbon and aggregation under poplar based agroforestry system in relation to tree age and soil type. *Agroforestry Systems*, 76(1), 27-35.

Kukal, S.S., K.L.Khera and M.S. Hadda (1993). Soil erosion management on arable lands of submontane Punjab, India: A review. *Arid Soil Research and Rehabilitation*, 7(4), 369.

Kukal, S., H.Sur and S. Gill (1991). Factors responsible for soil erosion hazard in submontane Punjab, India. *Soil Use and Management*, 7(1), 38-44.

Lal, R. (1989). Agroforestry systems and soil surface management of a tropical alfisol: II: Water runoff, soil erosion, and nutrient loss. *Agroforestry Systems*, 8(2), 97-111.

Lal, R. (1990). Agroforestry systems to control erosion on arable tropical steeplands. In Research Needs and Applications to Reduce Erosion and Sedimentation in Tropical Steeplands (pp. 338-345). Fiji: IAHS-AISH Publ. No.192.

Le Bissonnais, Y. (1996). Soil characteristics and aggregate stability. In M. Agassi (Ed.), Soil erosion, conservation, and rehabilitation (pp. 41–60). New York: Marcel Dekker Inc.

Morgan, R.P.C. (1995). Soil erosion and conservation (2nd ed.). UK: Longman Group Limited.

Narain, P., R. K.Singh, N.S. Sindhwal and P. Joshie (1997). Agroforestry for soil and water conservation in the western Himalayan Valley Region of India 1. Runoff, soil and nutrient losses. *Agroforestry Systems*, 39(2), 175-189.

Rashid, G., and S.Arora (2007). Physiography in relation to sustainable agriculture in Jammu. In S. Arora, S.S. Kukal, & V. Sharma (Eds.), Natural resource management for sustainable hill agriculture (pp. 93-100). Jammu, India: Soil Conservation Society of India, Jammu Chapter.

Saha, D., S.S. Kukal and S. Sharma (2010). Landuse impacts on SOC fractions and aggregate stability in typic ustochrepts of Northwest India. *Plant and Soil*, 339(1-2), 457-470.

Sharda, V.N. (2005). Integrated Watershed Management: Managing Valleys and Hills in the Himalayas. In B. R. Sharma, J. S. Samra, C. A. Scott, & S. P. Wani (Eds.), Watershed Management Challenges: Improving Productivity, Resources and Livelihoods (pp. 61-81). Colombo, Sri Lanka: International Water Management Institute.

Sharma, P.D. (2004). Managing natural resources in Indian Himalayas. *Journal of the Indian Society of Soil Science*, 52(4), 314-331.

Sharma, U.C., and V. Sharma (2003). Socio-economic aspects of tillage and its effect on soil erosion and agricultural productivity in Kandi (Drought prone) region of Jammu. In Proc 16th Int. soil Tillage Res. Org. Conference on Soil Management for Sustainability (pp. 1108-1113). Brisbane, Austraia.

Sharma, V., S.Arora and V.K. Jalali (2004). Studies on constraints in adoption of improved soil and rainwater conservation practices in kandi region of Jammu. *Journal of Research, SKUAST-J*, 3(2), 214-220.

Sharma, V., S. Arora and V.K. Jalali (2005). Rainwater Management on Micro-Watershed Basis in the Kandi region of Jammu. In V.N. Sharda, K.S. Dhadwal, S.K. Dhyani, G.P. Juyal, P. Dogra, A. Raizada, & O.P.S. Khola (Eds.), Resource Conserving Technologies for Social Upliftment (pp. 378-382). Dehradun, India: IASWC.

Sharma, V., S.H. Mir and S. Arora (2009). Assessment of fertility status of erosion prone soils of Jammu Siwaliks. *Journal of Soil and Water Conservation*, 8(1), 37-41.

Sidhu, G.S., K.P.C. Rana, T. Lal, S.K. Mahapatra, T.P.Verma and R.V.S. Rao (2007). Soils of Himachal Pradesh: Land capability classification and assessment of soil degradation status for suggested land use. *Journal of the Indian Society of Soil Science*, 55(3), 335–339.

Singh, M.J. (2010). Pattern and behaviour of gully erosion in Shivaliks of lower Himalayas. In Soil Solutions for a Changing World (pp. 98-100). Presented at the 19th World Congress of Soil Science, Brisbane, Australia.

Singh, M.J. and K.L. Khera (2009). Physical Indicators of Soil Quality in Relation to Soil Erodibility Under Different Land Uses. *Arid Land Research and Management*, 23(2), 152. doi:10.1080/15324980902817147

Singh, M.J. and K.L. Khera (2007). Soil erodibility in relation to physical and physico-chemical characteristics of some submontane soils of Punjab. *Journal of the Indian Society of Soil Science*, 55(3), 340-348.

Singh, M.J., and K.L. Khera (2008). Soil erodibility indices under different land uses in lower Shivaliks. *Tropical Ecology*, 49(2), 113–119.

Wischmeier, W.H. and J.V. Mannering (1969). Relation of Soil Properties to its Erodibility. *Soil Science Society of America Journal*, 33(1), 131.

Young, A. (1989). Agroforestry for Soil Conservation. CAB International ICRAF.

Chapter-15

Bamboo Species of Shivalik Foot Hills

D.R. BHARDWAJ, KRISHAN CHAND AND BHALENDER SINGH
Dept. of Silviculture and Agroforestry, College of Forestry, Dr. Yashwant Singh Parmar University of Horticulture & Forestry, Nauni, Solan (H.P.) 173230

ABSTRACT

The paper discusses the problems of soil and water erosion in the Shivalik foot hills and suggests strategies for improved land use planning and the role bamboo species can play in mitigating the problems. The paper also discusses the various bamboo species of the Shivalik foothills, their regeneration, planting practices, management and utilization pattern in detail.

Introduction

The Himalayan region in India is divided into four distinct zones according to altitude namely the Shivaliks (900-1,500 m), the outer Himalayas (1,500-3,600 m), the Middle Himalayas (3,600-4,600 m) and the greater Himalayas (above 6,100 m) (Sharma, 2008). The Shivalik region in north India stretches over a length of about 400 km, covering 8 million ha of foothills ecosystem and lies between 30° 10′ to 33° 37′ N° latitude and 73°37′ to 77°39′ E longitude. It has been identified as one of the eight most degraded rainfed agro-ecosystems of the country and hence included in the priority areas for watershed development. The rate of soil erosion in Shivalik is greater as compared to other physiographic regions of the Himalayas. Greater magnitude of soil loss in the Shivalik has been linked to immature geology and high degree of weathering of rocks. Soils are formed from river borne deposits derived from the rising Himalayas. The landscape is typically of low rolling hills, bisected by innumerable gullies, seasonal streams, commonly termed as *choes*, which drain the region. Today, most of the Shivalik region is a scene of dusty, rolling topography and devoid of proper vegetative cover. Monsoon torrents, numbering in hundreds, play havoc during monsoon. They cut across fertile fields and spread sheets of sand and gravel over the flooded area, turning them into wasteland. More than 80 t $ha^{-1}yr^{-1}$ soil has been recorded to be lost from denuded hills of Shivalik's due to soil erosion. Such a high rate of degradation needs to be checked through implementation of wasteland management programmes (Tiwari, 2008).

Agriculture, in Shivalik region is of subsistence type. Shortage of fodder and fuel wood caused by deforestation, decreased vegetative cover and severe soil erosion are some of the problems of the region. The local people and their relatively large cattle population depend on the forests situated in the Shivalik hills for their day-to-day needs and in the process has reduced this natural resource to a dismally low level resulting in lowering of economic and environmental services provided by it. Declining soil productivity and loss of forest based natural resources in the region is believed to be the major reason for backwardness of the people of Shivaliks. Therefore, reversing the degradation of natural resources is the key to promoting development in Shivalik region.

Strategies for improved land use planning

In the background of the problems and potentials identified and discussed as above, the following points should be considered for development of the region.

- Man made ecological degradation should be reversed specially in the lower foothills through land development activities like- land shaping, terracing, bunding, revegetation, etc;
- Rehabilitation of higher hills through reduced runoff and erosion, especially in shallow, sandy and skeletal soils on steep slopes;
- It can be done by adopting suitable soil and water conservation measure and afforestation/vegetation activities in the form of

planting tree species like bamboo, Khair, Siris, Chhal, Toona, Khirk, Biul, Bhahera, Subabul, Arjuna and grasses, etc. Plantation of these species should be encouraged to check soil erosion.

- Out of the above mentioned tree species, bamboos have got the potential of offering solutions to the wide range of problems being faced by people in the foothills of Shivalik region.

Why bamboo needed for Shivalik foot hill slopes?

- It slows down the rate of deforestation. Fast growing bamboo will quickly replenish a forest stripped of vegetation.
- Bamboo anchor topsoil along steep slopes and riverbanks, therefore very effective in controlling soil erosion.
- It improves the hydrological potential of site and promote soil health. Some species of bamboo absorb as much as 12 tones of CO_2, per hectare per year basis. Therefore, a valuable asset to deploy against global warming.
- It creates a perfect/suitable habitat for wildlife.
- It is an additional cash crop to the farming community.
- Bamboo shoots offer an important source of nutrition to the poor people living in the harsh climate.
- Bamboos are some of the few species that survive cyclone, where many other species get uprooted.
- Bamboo species have been found to clean the environment by removing toxins and heavy metals from dirty water, which is quite common in Shivalik foothills because of heavy concentration of the industries in this region.

There is an unfulfilled potential for bamboo to protect forests and agriculture, stream and ponds bank stabilization.

Bamboo species for Shivalik foothills

The bamboo species which grows or cultivated in the shivalik foot hills are *Dendrocalamus strictus, Bambusa arundinacea, Bambusa nutans* and *Dendrocalamus hookeri* var. parishii. Out of these bamboo species only *Dendrocalamus strictus* form dry bamboo brakes in the natural Tropical dry deciduous forest (5B/E_9). Whereas, all other bamboo species are cultivated or reared by the farming community on their private land for their multifarious uses.

1. *Bambusa arundinacea* Syn. *Bambusa bambos* (Spiny bamboo)

The species grows well on flat alluvial and moist soil. It prefers moderately fine clayey soil and tolerates slight water logging and avoids coarse grained soils and open ridges. This species is widely cultivated in Sub-Himalayan and Indogangetic plain.

2. *Bambusa nutans* (*Nal bans*)

It grows well on rich loamy soil and flourishes best in moist localities along river banks, depression and nallah in valleys. It is naturally occurring in sub-Himalayan tract from Yamuna eastwards to Arunachal Pradesh between 600-1500 m. It is commonly cultivated by the villagers in their homestead in Kangra, Hamirpur, Bilaspur, Una and some parts of Solan, Mandi, Chamba and Sirmour districts.

3. *Dendrocalamus hamiltonii* (*Maggar bans*)

It is capable of growing on wide variety of soils having good drainage, but performs best on sandy loam soil underlain by boulders along streams, valley and flat lands. This species is indigenous to Central and eastern Himalayas and perform well upto an elevation of 1,000m in moist deciduous and evergreen forests. It is extensively cultivated in Himachal Pradesh between 350-1,800m asl altitude in the districts of Kangra, Hamirpur, Bilaspur and in parts of Chamba, Una, Shimla, Solan and Sirmour having a rainfall of 800 m annually.

4. *Dendrocalamus strictus* (*Lathi bans*)

It grows on hill slopes on well drained dry soils and avoids poorly drained moist heavy soils. It grows equally well in moist as well as in dry areas. It occurs on hilly tracts usually upto an altitude of 1,000m. It is most common bamboo species of Shivalik region.

5. *Dendrocalamus parishii* (*Faglu*)

This species grows on rich loamy soil in valleys and along streams in moist locality. It is sparsely distributed in Shivalik foothills in the district of Hamirpur, Una and Kangra.

Regeneration

Bamboos can be regenerated by natural as well as artificial means.

(A) Natural Regeneration

Bamboo can be naturally regenerated by seeds and vegetative parts (Rhizome). Profuse natural regeneration of bamboos occurs after a gregarious flowering in *Dendrocalamus strictus* growing in the Shivalik foot hill conditions. Seed germination starts with the commencement of rainy season. After which, a large number of seedlings appears in the vicinity of the flowered clumps, which lead to severe competition among themselves as well as with the other species of the area. These seedlings have to face a severe biotic and abiotic pressure during the first year. Fire, grazing, repeated hacking of young culms, trampling of seedlings by cattle and monkeys cause extensive damage to the young seedlings. Those seedlings, which survive during the first year, take the place of flowered clumps after 8-10 year interval. Young seedling do not relish heavy growth of weeds, high light tree shade is helpful in protecting them from drought and frost. The process of natural thinning of seedlings helps in establishment of clumps at an optimum spacing from one another. Under Himachal Pradesh conditions natural regeneration of *D hamiltonii* is almost nil. In this species, although the flowered clump springs up number of seedling in its neighbourhood but all of them die during the ensuing dry and cold periods of autumn and winter seasons, respectively.

Bamboo in between two seeding periods reproduces itself through rhizomes. Each rhizome has got the potential to give rise to culm. Rhizomes are activated with the increase in the soil temperature and starts coming out of the soil in the form of culm with the onset of rainy season. It reaches full dimension in 60 to 120 days. The young culms grow to its full height before branching.

(B) Artificial Regeneration

Bamboos can be regenerated artificially by directly sowing seeds, planting out nursery-raised seedlings and by vegetative methods. Though, there are a number of advantages in raising bamboos through seeds, but due to irregular availability of seeds and their short viability in storage, bamboos are generally raised by vegetative methods. Vegetative propagation methods are used for the production of planting stock, when seeds are not available and/or genetically superior stock is to be multiplied. Two major types viz., macro and micro propagation methods are used in vegetative propagation of bamboos. In macro-propagation, conventional methods of offset planting, rooting of cuttings (culm and branch) and layering are used while in micro-propagation tissue culture techniques are being employed.

The propagation techniques of bamboo species (Bhardwaj and Mishra, 2006) growing in Himachal Pradesh is presented in Table 1.

Table 1 : Propagation techniques of bamboos of Shivalik

Species	Propagation techniques
Bambusa bamboos	Seeds, macro-proliferation, culm segments & offsets, etc.
Bambusa nutans	Seeds, macro-proliferation, offsets, culm cuttings, etc.
Dendrocalamus hamiltonii	Seeds, macro-proliferation, offsets, culm cuttings and branch cuttings, etc.
Dendrocalamus strictus	Seeds, macro-proliferation, offsets, culm cuttings, etc.
Dendrocalamus hookeri var. parishii	Seeds, macro-proliferation, offsets and branch cuttings, etc.

Planting practices

Bamboos have a relatively wide tolerance of site conditions. However, it is expected to show vigorous growth under conditions of high soil fertility, good drainage and abundant rainfall, which is well distributed throughout the year. The bamboo species growing under Shivalik foothills prefer almost year round high temperature. They generally thrive well on drained sandy loam to clay loam soil having pH of 5 to 6.5 but grow poorly on too acidic, saline-sodic and coastal salty soils.

The original vegetation on the land and the goal of plantation determines site preparation. If the purpose is to develop commercial bamboo stands then the original shrubs and herbaceous plants need to be removed, leaving only the trees. If the stands are developed for ecological amelioration, clearance of vegetation around the pits will be enough. The pits should be dug out at least one month before planting and must be large enough. Normally the pit size varies from 30 cm^3 to 45 cm^3, which may further be altered depending upon the size of the propagule and site conditions. Adequate quantity of well decomposed farmyard manure or compost 3-5 kg per plant should be mixed with the soil during the refilling of pits. The application of anti-termite insecticides is found essential in termite prone areas. The plant spacing/ distance should be delimited according to normal size of clump and site productivity usually ranging 4-9 m.

Seedlings can be transported using any appropriate means, considering the distance from the nursery and site access. The seedlings must be handled carefully to avoid injuries to them. Naked rooted propagules should be wrapped in wet gunny bags or grass mats before transportation so as to retain moisture and avoid direct sunlight. Once lifted from the nurseries, the propagules should be planted in the fields as soon as possible. The containerized seedlings may be transported with their boxes or baskets to facilitate loading and unloading and to protect the seedlings from injuries. After planting, the area should be frequently weeded. Similarly, seedlings raised in the polybag should be transported in such a manner to avoid damage from shaking and jolting. Casualties are replaced in the year of planting and again at the beginning of next year monsoon.

Seed origin seedlings or macro-proliferated seedlings can be used for large-scale plantation programmes. The planting stock of culm origin and/or offsets is mainly used for raising homestead plantation. Planting should be done as soon as the rains have set in and should be completed by 15th of July to ensure better survival and vigorous growth. The planting distance should be according to the normal size of the clump. The site having low productive potential, the spacing adopted is less. Whereas, for productive site/homestead the planting distance should be more.

Management of young bamboo stands

In early stages, measure such as weeding and climber control is required to help seedlings establishment. Fertilization may be done some time after planting, if situation require. Inorganic fertilizers will be more suitable for the vegetative growth of bamboo seedlings. The following prescription may be used for every hectares of bamboo stands; Nitrogen fertilizer: 20-30 kg; Phosphorus fertilizer: 10-15 kg; Potassium fertilizer: 10-15 kg in two splits. The best time for the first fertilizer application can be one month after planting, when the transplanted seedlings have their growth restored. The second fertilizing may be performed 4 months later. Fertilizer application may be advanced or postponed according to the growth state of the bamboo plants. Bamboo forests usually have a greater tolerance to fire than other forests. When forest fires consume the above ground parts, the underground rhizome can easily sprout and shoots will emerge in the following year to form a new stand. Nevertheless, efforts should be made to prevent fires, as these are very destructive to bamboo forests and pests and diseases can easily infect burnt stand. Besides, burning may stimulate the bamboo flowering, which can cause the death of large area of forest stands.

The litter in and around bamboo stands-such as fallen leaves, dry twigs, dead shoots are highly influenced and often makes the bamboo stand an area susceptible to fire hazard. Firebreaks and greenbelts are possible preventive measures in bamboo stands. Besides fire prevention, the greenbelt can also act as a windbreak. Site-specific firebreaks or greenbelts should be designed at the planning stage. Strict fire prevention measure should be enforced in the gregariously flowered bamboo stands. As these stands are highly inflammable and thus prone to forest fire, which can destroy the emerging seedlings.

Grazing should be prohibited in young bamboo stands. Cattle will destroy the bamboo shoots and new leaves. Rodents have the tendency to feed on bamboo shoots, make holes at the basal part of the bamboo stem and eat up the roots and the rhizome. Rodents attack the bamboo plantation in big way during winter months (December-January) under dry weather conditions. Ineffective rodent control frequently leads to degradation and even destruction of bamboo stands. Bamboo plantation established in or near *Lantana* infested areas suffers badly from depredation by porcupine, rats and pigs. In some situation young bamboo are also affected largely by fungus as well as insect-pest. Integrated pest management measures should be adopted, especially upto 2 years.

Utilization

Bamboo is one of the most important renewable resources gifted to the mankind. The strength of bamboo culms, their straightness, and lightness combined with hardness, range and size, abundance, relatively easy propagation and the short period in which they attain maturity, make them suitable for variety of uses. Practically all parts of bamboo i.e. leaves, young shoots, culms, rhizome and seeds are used in various ways.

A brief account of the utilization pattern (Bhardwaj and Mishra, 2006) of the different bamboo species occurring in Shivalik hill conditions is presented in Table 2.

Table 2 : Economic uses of bamboo species occurring in Himachal Pradesh

Species	Utility
Bambusa bamboos	Paper industry, thatching and roofing, basket making, bows and arrows, mats, agricultural implements, seed food, floats for timber, rafts, fencing, shooli for food, sericulture industries, chicks for door and window, general utility, horticultural pursuits, tomato sticks, cremation, coffins, cradles, scaffolding, cart yokes, ladders, protection during grain pounding, country tiles, cat sheds and roofs, seed drills, containers for administering medicines to animals, boat roofs, trellises, etc
Bambusa nutans	Construction, Basket making and fodder, etc.
Dendrocalamus hamiltonii	Construction, float for timber, rafts, mats, Basket making, shooli for food, cremation and fodder, etc
Dendrocalamus strictus	Paper industry, construction, walking sticks, thatching and roofing, basket making, bows and arrows, mats, agricultural implements, fodder, tool handles, fishing rods, seed food, floats for timber, rafts, fencing, shooli for food, sericulture industries, chicks for door and window, general utility, haystack stabilizers, horticultural pursuits, cremation, coffins, cradles, scaffolding, cart yokes, ladders, protection during grain pounding, country tiles, cat sheds & roofs, seed drills, containers for administering medicines to animals, boat roofs and seed drills, etc
Dendrocalamus hookeri var. parishii	Construction, paper, basket making and fodder, etc.

All bamboo species are suitable for afforestation of riverbanks, soil conservation, shelterbelts and as wind breaker, etc.

References

Bhardwaj, D.R and V.K. Mishra (2006). Bamboos of Himachal Pradesh, Y.S. Parmar University of Horticulture and Forestry, Solan, 33pp.

Sharma, P.K. (2008). Water management for agriculture in Shivalik regions. *In*: *Wasteland in Himalaya: The Shivalik*; Edited by D. Tripathi, R Bava and JN Raina. pp.97-111.

Tiwari, A.K. (2008). Natural resource Management in Shivaliks. *In*: *Wasteland in Himalaya*: *The Shivaliks*; Eds.D. Tripathi, R Bawa and J.N. Raina. pp.13-14.

Chapter-16

Effect of Different Land Treatments on Hydrological Behaviour and Soil Loss in Shivaliks

V.K. BHATT AND A.K. TIWARI
Central Soil and Water Conservation Research and Training Institute
Research Centre, Chandigarh-160 019

ABSTRACT

The Shivalik range of Himalayas, represent the most fragile ecosystem in India. In order to see the impact of land treatment, five hilly micro watersheds typical of the region having catchment area from 0.813 to 4.750 ha were selected for study. All the five watersheds were brought under study during 1966. Temporal changes in water yield from hilly catchments of Shivalik were monitored for twenty eight years (1966-1993). The watersheds were treated with annual burning (LT1), contour trenching +afforestation (LT2), cutting of trees, shrubs and over grazing (LT3) and over grazing alone (LT4). One watershed (LT5) was kept as control without any interference. As a result of burning, cutting and grazing there was significant increase in runoff as well as peak flow in the initial years, however, after increase in vegetation through protection from 1977 onwards resulted in decrease in runoff and soil loss.

Introduction

Among all products obtained from forests, water is most important for survival of mankind. Similarly, stream flow from forests supply considerable water for drinking and irrigation purposes. Removal of forest cover accelerates the rate and volume of runoff. In some areas, cutting trees causes a temporary increase in volume of water flowing downstream. At some places there is political pressure for cutting trees in order to increase the water especially in Asian countries due to rise in population. However cutting of trees for increasing water yield is not sustainable as the increase in flow rate and volume is for a short period and the practice can ultimately degrade water quality and increase vulnerability to flooding. The study of water flows from forest can help to understand the relation between forest and water. But now it has to deal with today's complexities, including climate change, forest fires and changing pattern of development and ownership.

Forest harvesting/burning of trees or other reductions to vegetative cover, generally increases the average surface runoff volume and total water yield for a given area of land. These land cover alterations decrease time of concentration of flow, increase the intensity of peak flows for a given precipitation event, and increase the frequency and intensity of extreme flow events, especially channel flows. These alterations tend to deteriorate water quality by transporting sediment and other pollutants from the landscape and increasing erosive forces within the stream channel. Studies have shown that afforestation (re-establishment of forest cover) results in a decrease in water yield, though responses to treatment are highly variable and unpredictable. Studies have also shown that afforestation produces a reduction in water yield proportionally to the growth rate of the stand. Attempting to quantify the impact of forest harvesting on annual water yield is time consuming and expensive. It requires a long-term commitment from researchers, which necessitates sufficient funding.

Hibbert (1967) reported for thirty-nine studies the effect of altering forest cover on water yield. Taken collectively, those studies reveal that forest reduction increases water yield, and that reforestation decreases water yield. Results of individual treatments vary widely and for the most part are unpredictable. First-year response to complete forest reduction varies from 34 mm to more than 450 mm of increased streamflow. A practical upper limit of yield increase appears to be about 4.5 mm per year for each per cent reduction in forest cover, but most treatments produce less than half this amount. There is strong evidence that in well-watered regions, at least, streamflow response is proportional to reduction in forest cover. As the forest regrows following treatment, increases in streamflow decline; the rate of decline varies between catchments, but appears to be related to the rate of forest recovery. Seasonal distribution of streamflow response to treatment is variable; response in streamflow may be almost immediate or considerably delayed, depending on climate, soils, topography, and other factors.

Moore and Wondzell (2005) from their study reported that forest harvesting generally increases the fraction of precipitation that is available to become stream flow, increases rates of snowmelt, and modifies the runoff pathways by which water flows to the stream channel. In small headwater catchments, forest harvesting generally increases annual runoff and peak flows and reduces the severity of low flows, but exceptions have been observed for each effect. Low flows appear to be more sensitive to transpiration from vegetation in the riparian zone than in the rest of the catchment.

Shivaliks

Shivalik region of India covers an area of 3.1 million hectares of which the maximum area falls in the state of Himachal Pradesh followed by Punjab, Jammu & Kashmir, Uttarakhand, Haryana and union territory of Chandigarh (Mittal *et al.*, 2000). These represent the most fragile ecosystem in India. High intensity rainfall with rolling topography induces large volume of runoff giving rise to various seasonal and perennial rivers. Moreover, over exploitation of ground water and high biotic pressure results in ground water table depletion and natural resource degradation.

The Shivaliks are characterized by low hills, undulating topography, steep slopes, erodible soils and are dissected by numerous seasonal streams. The elevation of shivalik range from 280 m to 1500 m. Average annual rainfall of the region is 1100 mm with about 82 per cent being received during monsoon season (July to September) with average annual pan evaporation value of 2110 mm. Temperature varies from 2° C to 43°C. In general soils are coarse textured, deep with pebbles and boulders on slopes near the foothills. The rain-fed areas of Shivaliks have no possibility of providing irrigation through conventional methods and large amount of runoff not only goes waste from such lands but also creates problem of soil erosion and land degradation. Management of water resources in its various forms is key to environmental and economical sustainability in the region.

Study Area

The selected watersheds are located at the Research Farm of CSWCRTI, Research Centre, Chandigarh, in Pinjore block of Panchkula district in Haryana state. They are located at elevation of about 450 m (above msl), at 30° 45′ N latitude and 76° 52′ E longitude.

The study was conducted on five natural, micro hilly watersheds located within the Research Farm of Centre. All the micro-watersheds were being gauged by 0.6m deep 2:1 broad crested triangular weirs. The runoff was being measured by automatic water level recorders. All the watersheds were calibrated for initial few years (1966-72). Areas of these micro-watersheds vary from 0.736 ha to 4.153 ha.

Vegetation

Owing to the past misuse of the land, tree growth was almost absent in all the watersheds in the initial years. The scrub growth consisted principally of *Carissa opaca* (Karanda) and *Ziziphus numularia* (Ber). There were few trees of *Acacia catechu* (Khair) and *Dalbergia sissoo* (Shisham). The grass cover included *Heteropogon contortus, Chrysopogon fulvus, Dicanthium annulatum, Eulaliopsis binata* and *Eucmopogon fovoolatus.*

Material and Methods

The standard characteristics of all the five watersheds under study have been worked out, based on detailed contour maps of the entire watersheds. The steep overland slope of these five watersheds, varying from 32.1 to 50.6 per cent represent the shivaliks of this region. In spite of such steep slopes and the south-western aspect, the prevailing ground cover has been fully effective in reducing surface runoff and soil loss to almost negligible. The standard characteristics are given in Table 1.

In order to determine the effect of different treatments on runoff and soil loss an experiment was laid out in Research Farm over an area of 11 ha. The watersheds were treated with annual burning (LT1), contour trenching + afforestation (LT2), cutting of trees, shrubs and over grazing (LT3) and over grazing alone (LT4). One watershed (LT5) was kept as control which means no interference from outside or inside. As a result of burning, cutting and grazing there was significant increase in runoff as well as peak flow. In order to establish basic hydrological relationships between the experimental watersheds, calibration studies were started during 1966. The soil type of experimental watershed is classified as Chandigarh series-1. Soils are deep calcareous through out loamy in surface and silty clay loam in sub soil. Colour varies from reddish brown to light brown. pH varies from 7.9 to 9. The average runoff percentage from all the watersheds during the period from 1966 to 1972 ranged from 11.3 to 24.0% of the annual rainfall.

Table 1 : Watershed Characteristics

Watershed Characteristics	Watershed Number				
	LT-1	LT-2	LT-3	LT-4	LT-5
Area, ha	4.482	2.020	4.750	0.813	1.695
Aspect	South West	South West	South West	South West	South West
Maximum length, m	365	245	355	175	215
Maximum width, m	160	115	155	65	125
Average width, m	99.6	79.2	113.1	42.8	70.6
Perimeter, m	982.5	575	925	410	570
Length of main drain, m	450	255	420	190	240
Length of total drain, m	1802	525	1925	392	866
Drainage Density, km/sq km	40.2	26.0	40.5	48.2	51.1
Slope,%	32.1	14.2	19.4	9.2	25.1
Compactness Coefficient $0.28P/\sqrt{A}$	1.30	1.13	1.19	1.27	1.23
Land use	Shrubs Grasses and scattered trees	Shrubs Grasses and scattered trees Affo-restation	Shrubs Grasses and scattered trees	Shrubs Grasses and scattered trees	Shrubs Grasses and scattered trees
Treatment imposed in 1973	Annual burning		Cutting of trees and shrubs/ over grazing	Overgrazing	Control

Results and Discussions

All the five watersheds were calibrated before applying the treatments in the year 1973. The calibration equations for the period 1966 to 1972 worked out for each watershed (LT1 to LT5) were as under:

$Y_1 = 0.056P+0.36$ (r = 0.88)

$Y_2 = 0.070P+0.055$ (r = 0.85)

$Y_3 = 0.049P+0.88$ (r = 0.0.83)

$Y_4 = 0.156P-2.9$ (r = 0.81)

$Y_5 = 0.094P+0.49$ (r = 0.87)

Where
Y = runoff (mm)
P = rainfall (mm)

Runoff from each of the four watersheds LT-1, LT-2, LT-3, LT-4 were calibrated with the runoff from LT-5 (Control).The equations for LT-1 to LT-4 are given as under:

$Y_1 = 0.68\ Y_5 - 0.30$ (r =0.78)

$Y_2 = 0.74\ Y_5 - 0.31$ (r =0.82)

$Y_3 = 0.52\ Y_5 + 0.63$(r = 0.80)

$Y_4 = 1.65\ Y_5 - 3.71$(r = 0.76)

*Where,*Yi is the storm runoff (mm) from i^{th} watershed.

With the help of above set of equations, expected runoff from each of the watersheds before treatment was calculated.

Investigations were conducted earlier during 1963 on the infiltration rates in the experimental watersheds. The average intake rates derived for the watersheds are as follows:

$Ir_1 = 10.8T^{-0.40}$

$Ir_2 = 5.0T^{-0.40}$

$Ir_3 = 2.83T^{-0.40}$

$Ir_4 = 5.60T^{-0.40}$

$Ir_5 = 4.23T^{-0.40}$

Where, Ir_1, Ir_2, Ir_3, Ir_4 and Ir_5 is intake rate in hours and T is time in hours.

These equations indicate that all the five experimental watersheds have high infiltration capacity which accounted for low runoff.

The annual rainfall and runoff from all the forest watersheds under different treatments for the period from 1966 to 1993 are presented in Fig.1. The average values for the calibration period from 1966 to 1973 are also shown in same figure. The average runoff percentage from all the watersheds during the period from 1966 to 1972 ranged from 11.3 to 24% of the annual rainfall. During the calibration period watershed LT3 gave lowest runoff. The results indicate a drastic reduction in runoff and sediment yields after imposition of treatments in all the watersheds but the impact was highest in LT1 watershed. In watershed LT1 runoff reduced from 14.66% to 0.33%. This clearly indicated higher impact of annual burning of vegetation in comparison to effect of other treatments. In grassland watershed (LT4) runoff reduced from 24 % to 1.2% after rehabilitation followed by overgrazing. As a result of protection since 1978, other two watersheds LT1 and LT3 which were earlier under abuse of burning and cutting have also been restored to hydrological responses similar to those of the watersheds under afforestation (LT2) and control (LT5). It appears that the 4 years rest period has brought all the watersheds to almost the same hydrological response level. It is worth noting that inspite of steep slope the runoff is less than 1.26% in these watersheds

The sediment yield from these watersheds was also measured since 1980 for the same period for which discharge measurement have been reported. The sediment trapped in 3m x 1.5m x 1m stilling tanks at the gauging stations was measured at the end of the season. The suspended sediment in the outflow from the watersheds was practically negligible. The total sediment yield in all the cases varied from 0.75 to 97 t/ha^{-1} in the initial years when treatment were imposed. However, it reduced drastically from 28 t/ha^{-1} to 0.025 t/ha^{-1} in watershed LT1 from 1974-76 to 1980-89 respectively (Fig.2).

Conclusion

This long-term hydrological experiment could provide information about the hydrological behaviour of hilly forest watersheds and the impact of vegetation cover and the treatment of the catchment. Land treatments are effective for bringing change in runoff and sediment yield from the catchments. Annual burning resulted in good vegetation in the watershed which further resulted in reduction of runoff and sediment yield. This is clearly seen that proper land treatment which enhances the vegetation cover is more effective in controlling runoff and sediment in Shivalik watersheds.

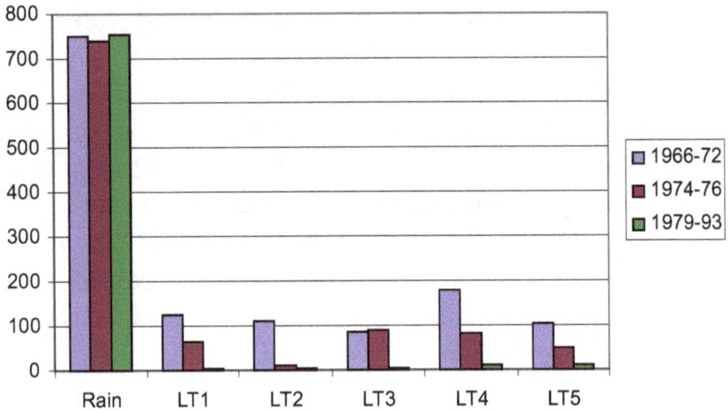

Fig.1 : Rainfall and runoff under different land treatments during different period

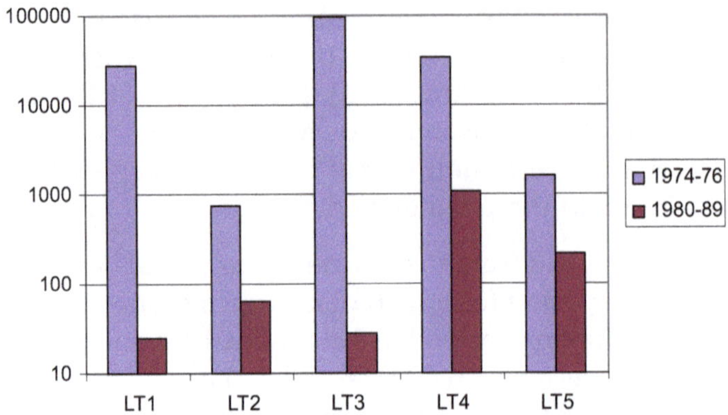

Fig. 2 : Sediment yield (kg ha^{-1}) under different land treatments during different period

References

Hibbert, A.R. (1967). Forest treatment effects on water yield. *In*: W. E. Sopper and H. W. Lull (Editors), *International Symposium for Hydrology*. Pergamon, Oxford, 813 pp.

Mittal, S.P., R.K. Aggarwal, and J.S. Samra, (2000). Fifty years of Research on Sustainable Resource Management in Shivaliks, CSWCRTI, RC, Chandigarh, 506pp.

Moore, R. Dan and S.M. Wondzell (2005). Physical hydrology and the effects of forest harvesting in the Pacific Northwest: a review *Journal of the American Water Resources Association*. 41(4): 763-784.

Chapter-17

Impact of Climate Change on Horticulture

W.S. DHILLON, SANJEEV K. CHAUHAN* AND H.V. SRINIDHI
Department of Horticulture, * Department of Forestry & Natural Resources
Punjab Agricultural University, Ludhiana-141 004

ABSTRACT

Among the challenges of 21st century, to make the food supply sustainable, the impact of climate change is largely unconceivable. The climate on earth has increased by 0.74° C over the past 100 years. Best estimate for low scenario of global warming is 1.8°C and for high scenario is 4°C. Last decade recorded the hottest in the century and 2009 has been rated hottest since 1900. Retreating glaciers have been confirmed in various parts of the world. Further, the current greatest challenge is to perceive the biological impact of such climate changes. Efforts have been initiated by Intergovernmental Panel on Climate Change (IPCC) to assess the impact of climate change on biological systems across the globe. Since 1950's, the change in carbon dioxide content in the atmosphere is being monitored. Almost a decade later, computer models were employed to assess the possible impact of change in temperature and enhanced CO_2. By the 1980's consensus was almost achieved on possible doubling of atmospheric CO_2 concentration due to emission from fossil fuels and rise in the global temperature. Since then quantitative estimates on the biological impacts of climate change have been taken up by scientific community world wide. Presently, ample of evidence on impact of climate change from marine to terrestrial environment, from species to communities, ecosystems and hierarchies have been gathered for both flora and fauna (Walther *et al.*, 2002). Parmesan and Yune (2003) reported that meta-analyses of 334 species and global analyses of 1,570 species revealed highly significant and regular patterns of change in relation to observed warming with more than 95 per cent confidence, as fingerprints of climate change.

Impact of Climate Change on Agriculture

The livelihood of more than seventy five per cent of Indian population depends upon agriculture. Still the climate change vulnerability and impact studies in India assume high degree of uncertainty in assessment due to limited understanding of many critical processes in the climate system. The costs of non addressing climate change or to adapt to it are very uncertain, but their consequences are enormous. Sathaye *et al.* (2006) have projected the impact of climate change and emphasized to take immediate remedial steps. The largest impact of climate change on agriculture will be due to the effects of induced changes in temperature, rainfall and sunlight. However, these effects are going to vary across the agro-climatic regions. The agricultural areas in Tropical Asia are vulnerable to many environmental hazards-including frequent floods, droughts, cyclones, and storm surges-that can damage life and property and severely reduce agricultural production. Shivaliks will be more prone to these hazards as it is more fragile then other ecosystems. The enhanced carbon dioxide leads to increased photosynthetic rates and leaf area development, which eventually translates to enhanced economic yield and biomass productivity (Krizek, 1986 and Allen, 1990). Various simulation techniques have been used to establish relationships between crop productivity and the projected climate change parameters. The simulation results indicated that increasing temperature and decreasing radiation levels pose a serious threat in decreasing crop productivity (Hundal and Kaur, 1996). Carbon dioxide fertilization is expected to favour the productivity and therefore will counteract the adverse effects of temperature rise in future. The doubling of the carbon dioxide level can increase the dry biomass to 10-15 per cent provided all other factors remain constant (Parry *et al.*, 1990). Central Research Institute for Dryland agriculture, Hyderabad has reported that the soybean yields could go up by 50 per cent on doubling the carbon dioxide in atmosphere. However, if this increase in carbon dioxide is accompanied by an increase in temperature, as expected, then the soybean yields could actually decrease. If the maximum and minimum temperatures go up by 1°C and 1.5°C, respectively, the gain in yield comes down to 35 per cent. If maximum and minimum temperature rises by 3°C and 3.5°C, respectively, then soybean yields will decrease by five per cent compared to 1998 level. Gregory *et al.* (1999) reported decrease in wheat yield as a consequence of warming and reductions in yields of rice of about five per cent per °C rise above 32°C. These effects of temperature were considered sufficiently detrimental that they would largely offset any increase in yield as a consequence of increased atmospheric carbon dioxide concentration. The potential consequences of changes in climate on the growth and yield of crop plants have been reviewed by many scientists, concluding that the earlier-anticipated benefits of carbon dioxide fertilization would be largely offset by nutrient limitations, pollutants and further interactions with climatic factors (Amthor, 2001; Fuhrer, 2003;Gregory *et al.*, 2005;Long *et al.*, 2005).

Agriculture will be adversely affected not only by an increase/ decrease in overall amount of rainfall, but also by shift in the time of the rainfall. In India, the crop area affected by floods is nearly one-third of the average flood-prone area (GOI, 1992) and in the neighboring Bangladesh about 0.5 million tons of rice is reported to be lost during 1962-1988 (Paul and Rashid, 1993). In eastern India, approximately 80% of the 20 million ha of rice is grown in rain-fed lowlands (Zeigler and Puckridge, 1995). More subtle changes in the groundwater level appears to be taking place in some high-productivity agricultural areas of the country (Sinha, 1997).

Moreover, the factors that increase the vulnerability of the agricultural sector in the region are the compelling features such a significant portion of agriculture is rainfed and a large human population depends on agriculture with 5.4 people per hectare in south Asia and 5.7 in Southeast Asia. In India, droughts during summer months are expected due to 5-25 per cent decline in winter time rainfall. Summer monsoon over the Indo-Gangetic plains has shifted westwards and onset of monsoon over different parts of the country may vary (Pandey, 2002). In Shivaliks the rainfall is erratic some time monsoon starts in June and some time in July. The intensity of the rainfall also varies which affects the vegetation and hence soil and water erosion. This invariably depends upon climate variation and hence also affects the climate. Anglo *et al.* (1996) opines that more vulnerable are low-income populations that depend on isolated agricultural systems, which have limited access to agricultural markets. Increasing population has also stressed agricultural production. During the past three decades, food production -largely rice and wheat-have been increased at manifold than the rate of population growth, however, currently these crops show trend of stagnation in productivity (Sinha, 1997).

Agricultural production will be sensitive to global climate change. The three major uncertainties regarding impacts of climate change on agriculture are, first, the magnitude of regional changes in temperature and precipitation, second, the magnitude of the beneficial effects of higher CO_2 on crop yields and third, the ability of farmers to adapt to climate change. Estimates of 1980s and early 1990s projected negative impact on agricultural production (Mendelson *et al.*, 1994), whereas, the recent studies show a range from slightly negative to slightly positive impact. Many developed nations in mid- to high-latitude locations, warmer temperatures may improve crop yields by extending the growing season. In contrast, many developing nations of sub-tropical and tropical areas are prone to global warming due to excessively high temperatures and reduced yields. Many crop models accord that the enhanced CO_2 increases yields of C_3 crops by about 20 - 35%, assuming a near optimum growth environment. However, field conditions are seldom optimum and this can be possible in situation where availability of water, fertilizer, and other inputs are adequate (Ruttan, 1999). Temperatures may become too hot, or be too low despite global warming, for beneficial effects of CO_2 in some areas. And with respect to C_4 crops, such as maize, sorghum, millet and sugarcane there would be very little benefit from higher CO_2, while these crops may face increased competition from C_3 weeds. Studies integrating climate, crop, and food trade models suggest that a moderate climate change may have only a small impact on world food production. The reduced yields in some parts of the globe are offset by increased yields in others. However, severe food shortages are likely to occur in some developing nations because of localized climate and resource constraints. Changes in soil, pests and weeds brought by climate change will also affect agriculture in India. The amount of moisture in the soil will be affected by changes in factors such as precipitation, runoff and evaporation.

Climate and Horticulture

Almost all horticultural plants are highly sensitive to the direct effects of rising atmospheric carbon dioxide (CO_2) concentrations, as well as temperature and precipitation. Climate change and CO_2 are likely to alter important interactions between horticultural plants and pollinators, insect and disease pests, and weeds. Being perennial, horticultural crops are more dependent on climate.

Temperate fruits

The temperate deciduous fruit crops which respond to seasonal changes are grown in temperate region of the country where the temperature falls below freezing point in winter. Dormancy and activeness are distinct with strong correlation with climate. During autumn, leaves are shed and plants become winter hardy. Growth hormones play an important role in phenology. Rest is dormancy situation when growth is arrested even under favourable climate. Winter chilling breaks the rest period. The chilling requirement of species decides its suitability for cultivation. Upper parts of the Shivaliks harbors some of the temperate species like plum and apricot. The chilling requirements of some common temperate fruits are given in Table 1.

Table 1: Minimum chilling requirements of temperate fruits

Fruit crops	Hours (< 7°C)
Pear	1500-1600
Bartlette	1500
Apple	1000-1600
Plum	700-1000
Almond	250-500
Peach	200-850
Apricot	700

Sub-tropical fruits

These fruits are grown in climate where a hot and dry summer with less severe winter prevails. There is a distinction between summer and winter and receives less rains. The temperature ranges from 25 to 30°C with high humidity during monsoon. Fruits such as citrus, fig, guava, grapes, pomegranate, phalsa and litchi are commonly cultivated in this region. About 80 per cent of the Shivalik Himalayas falls under this group. In citrus, high temperature during fruit maturation period results in poor pigmentation leading to poor color of fruits and also cause sun burns. Citrus is very sensitive to low temperature and cannot survive if temperature falls below -9°C. In grapes, rainfall quantity, duration and season influences the crop much. Rainfall during flowering and fruiting is detrimental. Increased humidity due to prolonged rainfall makes fruits tasteless and causes cracking of skin. Flower and fruit drops due to high wind. In litchi, humidity is an important factor influencing fruit development. Dry hot wind in summer cracks the fruit and pulp gets damaged. The area should be free from frost. The rise in temperature even by 1°C to 2°C during flowering and fruit set adversely affect the crop potential. Climate induced regulation of flowering is best revealed in pomegranate. Under tropical climate, pomegranate flowers throughout the year. In sub-tropical climate of central and western India, it flowers in 3 distinct phase, Ambe Bahar (January-February flowering), Mrig Bahar (June-July) and Hasth Bahar (October). In Karnataka, two distinct flowering during June-August and October-March is observed. In Punjab, only one flowering season is observed from April-June. In temperate climate, the crop flower during middle of April. Some fruit crops such as mango can grow better under semi-tropical and tropical climate. High temperature after fruit set improves fruit size and quality. Low temperature and frost during flowering limits better cropping. In Shivaliks, the production of fruits like citruses, *Embilica officinalis*, guava, plum, apricot will be affected due to various reasons associated with climate change. In the recent past it had been observed that during flowering period in April- May either temperature goes very high which desiccates the flowering buds, or sometimes there is hail storm during this time as in the year 2010. Thus leading to decrease is fruit quality and quantity.

Tropical fruits

There is no distinct summer and winter. Climate is characterized by humid summer and mild winter. The temperature fluctuation is relatively less. This climate prevails between 23.5°N and 23.5°S latitudes with little difference in temperature throughout the year. Mean monthly temperature is 18°C. The minimum heat requirements of the crops need to be satisfied for the flowering and fruiting (Table 2). Banana, mango, guava, pineapple, papaya, cashewnut, pineapple, etc. are commercially cultivated. The sensitivity of fruit crops to climate can be better noticed in banana. Being a tropical crop, temperature is known to control growth, development and productivity (Chattopadhyay, 2006). Temperature around 25°C is conducive for growth and fruit matures steadily whereas, temperature beyond 32°C is detrimental due to higher photorespiration. Coconut is another fruit which is exclusive for tropical conditions. It cannot tolerate long spell of drynesss and severe winter. It fails to grow in climate which experience below 27°C. In India, it is restricted within 26°N latitude and altitude up to 1000 m above sea level. The pine apple is a crop of tropical climate but can thrive in sub tropical conditions. The region must be frost free and cannot tolerate high temperature as well. Under such conditions acid content of the fruit increases and quality decrease.

Table 2: Heat requirements of tropical fruits (from flowering to first picking)

Fruit	Varieties	Heat Units (day °C) (above 16°C)
Apricot	Early	1155-1245
	Mid-late	1725-2186
Banana	-	2144
Grapes	Perlette	1051.1
	Beauty seedless	967.4
	Angur early	890.10
	Arka Shyam	1491.7
	Arka Hans	1418.7
	Cardinal	1117.20
Datepalm	Early	1600
	Late	3000
Pecanut	-	5000

Impact of Climate Change on Crop Physiology

The events such as leaf flush, flowering, fruit set and fruit quality strongly depends on climatic conditions and these events directly influence the commercial cultivation of the fruits. Recent studies have documented that during the past several decades, significant shifts in plant phenology and insect migration patterns have occurred for some species in the U.S. and Europe (Walther *et al.*, 2002; Menzel and Eastrella, 2006).

In temperate regions of India, apple is largest grown commercial fruit crop. The decadal studies on availability of chilling requirements and growing degree hours in Mashobra region of Himachal Pradesh revealed that duration of rest period is negatively correlated with fruit production (Table 3). Chilling units had positive effect. Intermittent warm period affected the rest requirement and manifested in poor cropping. In parts of Solan (Himachal Pradesh) which falls in Shivalik Himalayas, it had been seen that the usual time of bud sprout in fruit plants had advanced because the average temperature of February and March months had increased.

Table 3: Correlation between climatic parameters and apple productivity in Himachal Pradesh

Parameters	r value
Rest period	-0.716**
Chilling period	+0.44*
Peak winters	+0.642**
Growing degree hours (GDH)	
- Chilling initiation to rest completion	-0.67**
- Rest completion to full bloom	+0.64**

*Significant at $p<0.05$; ** ($p<0.01$)

Effect of drought on fruit set, yield and quality of custard apple was studied by George and Nissen (2002). The study revealed that severe drought negatively affected the fruit size. This was attributed to effects of heat stress on cell division after fruit set.

Effect of high temperature

Most plant processes related to growth and yield are highly temperature dependent and there can be a optimum temperature range for maximum yield. The optimum growth temperature corresponds to the optimum temperature for photosynthesis and other metabolic reactions. Higher temperatures speed up plant growth and development in annual crops. However, in perennial crops, being grown in a climate near its optimum, a temperature increase of several degrees could reduce photosynthesis and shorten the growing period affecting the productivity. The particular crop is currently being grown in areas where they are usually best-adapted to the prevailing climate. A significant increase in growing season temperature will require shifts to new varieties that are more heat tolerant and have a higher temperature optimum for photosynthesis. Developing such varieties is not an easy task where there are several inherent limitations to accomplish this. In many cases traditional fruit crops will have to be discarded for newer one better suited to the new environment. In areas where current temperatures are below optimal for specific crops, there will be a benefit, while in areas where plants are near the top of their optimal range, yields will decrease. In very cool regions, where the current climate limits their crop options, global warming will be mostly a benefit, giving them the opportunity to grow a wider range of crops and long-growing-season, high-yielding varieties. Some plant species require a cold period before they will produce flowers and a harvestable product. This process, called vernalization, has very narrow temperature and duration boundaries. Even a minor climate shift of 1-2°C could have a substantial impact on the geographic range of these crops. As fruit crops are perennial moving production areas is difficult.

Carbon dioxide effects

It is accepted fact that the atmospheric CO_2 is increasing exponentially and will likely double, about 700 parts per million within the next century (Wuethrich, 2000) unless some stringent measures are taken. This has a beneficial effect on the plants because plants utilize CO_2 during photosynthesis to produce sugars and energy for growth and development. It could also lead to significant increases in crop productivity. However, this is not universal as the biochemistry of photosynthesis differs among plant species. The C_3 plants benefit much more from increase in CO_2 than C_4 plants. However, the beneficial effects decline with long-term exposure to high CO_2. The greatest benefits from CO_2 tend to occur when plants are able to expand their sink capacity for the products of photosynthesis by producing more flowers and fruit when grown at high CO_2. But, the genetic or environmental factors limit growth and sink capacity. These leads to build up of sugars in the leaves, giving a negative feedback on photosynthesis and benefits from elevated CO_2 become minimal. The recent evidence suggests that photosynthesis is constrained by the activity of Rubisco enzyme under both high temperature and CO_2 (Jensen, 2000).

CO_2 and crop water use

Another important direct effect of high CO_2 on plants (both C_3 and C_4 species) is a partial closure of the small pores, or stomata, of the leaves. This mechanism restricts the transpiration more than it restricts photosynthesis. Some have suggested that this will moderate the increase in crop water requirements anticipated to occur in a global warming scenario. However, the plants grown at high CO_2 transpire less water per unit leaf area and their leaves are larger with more stomata so that the whole plant water use is similar to or greater than plants grown at normal CO_2 concentrations. Several studies have found that the relative beneficial effect of a CO_2 doubling on growth is greater under optimum water supply.

The specific temperature range for a positive CO_2 effect varies and for most crops the beneficial effects become minimal at temperatures below about 15°C. This has important implications for temperate regions where, the average temperatures during early and late portions of the growing season will be too cool. In this situation it is hard to expect much benefit from a CO_2 doubling. Under climatic high temperatures regimes, flowering and pollination of many crop species will be impaired. Yields will be very low regardless of atmospheric CO_2 levels.

Plant Health

Many researches examining the effect of elevated CO_2 on insect damage has found that leaf-feeding insects often must consume more foliage to survive. Because, the plant under high CO_2 regime tend to have a lower protein concentration. Natural selection would favor the evolution of insects that consume more plant material more rapidly. To combat this, more pesticides need to be used. Moreover, climate changes will influence the geographic range of insect-pests and diseases. Warmer temperatures in high latitude areas may give more duration for over-winter for insects. Diseases are more likely to increase in temperate regions because many fungal and bacterial diseases have a greater potential to reach severe levels when temperatures are warmer or when precipitation increases. The recent epidemic of citrus greening has devastated the citrus industry in Florida. The industry is reported to experience the loss of about $ 9 billion and 12.7 million trees have been cut as sanitary measure (Strokes, 2006). This was previously unknown in US, which is bacterial disease transmitted by psyllids. The trees infected do not show symptom for several years which makes difficult in early prediction of the losses.

Mitigation Strategies

Farmers in developing nations will be least able to adapt to climate change because of a relatively weak agricultural research base, poor availability of inputs such as water and seed of new varieties, and inadequate capital for making adjustments at both the farm and national level. Adapting to climate change will be costly as farming cost includes increased use of water, fertilizer, pesticides, etc. to maximize harvest under the higher CO_2 regime. In case of horticulture, the investment will be in taking new plantations and storage facilities to shifts/adapt to new crop/varieties. Environmental degradation associated with agriculture such as increased soil erosion, increased risk of ground and surface water pollution, depletion of water resources may emerge. The agricultural economies will collapse due to factors such as excessively high temperatures, severe pest pressure, lack of locally adapted varieties or poor markets. As climatic zones shift, there will be situation where there may be best suited climate for the crops but lacks good soils or available water for good harvest.

Carbon sequestration in the system is another strategy to fix the carbon through photosynthesis, which is thought to be one of the most cost-effective means of reducing atmospheric levels of CO_2. It is important to protect the ecosystems that store carbon and enhance the ability of the ecosystems to increase carbon sequestration beyond current conditions. The fruit breeders and physiologists must insure that the cultivars developed for economics and biological yield (above and below ground) should have elevated lignin and suberin contents that promote accumulation of stable carbon in plant products and soil.

References

Allen, L.H. (1990). Plant response to rising CO_2 and potential interactions with air pollutants. *J. Environ. Qual.* 19: 15-34.

Amthor, J.S. (2001). Effects of atmospheric CO_2 concentration on wheat yield. *Field Crops Res.* 73: 1-34.

Anglo, E.G., W.C. Bolhofer, L Erda., S. Huq, S. Lenhart, S.K. Mukherjee, J.Smith and J. Wisniewski (1996). Regional Workshop on Climate Change

Vulnerability and Adaptation in Asia and the Pacific: Workshop Summary. January 15-19, Manila, Philippines, Kluwer Academic Publishers, Dordrecht, The Netherlands, pp. 16.

Chattopadhyay, T.K. (2006). *The Text Book on Pomology (Tropical Fruits)* Vol.II. Kalyani Publishers, India. pp. 334.

Fuhrer, J. (2003). Agroecosystem responses to combinations of elevated carbon dioxide ozone and global climate xhange. *Agric. Ecosyst. Environ.* 97: 1-20.

GOI (Government of India), (1992). Eighth Five Year Plan (1992-1997). Planning Commission, New Delhi, India, pp 480.

George, A.P. and R.J. Nissen (2002). Effects of drought on fruit set, yield and quality of custard apple (*Annona* spp. Hybrid) 'African Pride' plants. *J Hort. Sci. & Biotech.* 77:418-427.

Gregory, P.J., J.S.I. Ingram and M.Brklacich, (1999). Managed production systems. In The terrestrial biosphere and global change:implications for natural and managed systems (Eds. Walker, B., Steffen, W., Canadell, J. and Ingram, J.S.I.) Cambridge Univ. Press, Cambridge UK, pp.229-270.

Gregory, P.J., J.S.I. Ingram and M. Brklacich (2005). Climate change and food security. *Phil. Trans. R. Soc. B.* 360: 2139-2148.

Hundal, S.S. and P. Kaur (1996). Climate change and its impact on crop productivity in Punjab, India. Climate Variability and Agriculture (Eds. Abrol, Y.P., Gadgil, S. and Pant, G.B.) Narosa Publishing House, New Delhi, pp. 377-393.

Jensen, R.G. (2000). Activation of Rubisco regulates photosynthesis at high temperature and CO_2. *Proc. Natl. Acad. Sci.*, 97:12937-12938.

Krizek, D.T. (1986). Photosynthesis, dry matter production and growth in CO_2 enriched atmosphere. Cotton Physiology No.1. The cotton foundation, Memphis, T.N., pp.193-225.

Long, S.P., E.A.Ainsworth, A.D.B. Leakey and P.B. Morgan (2005). Global food insecurity. Treatment of major food crops with elevated carbon dioxide or ozone under large-scale fully open-air conditions suggest recent models may have over estimated future yields. *Phil. Trans. R. Soc. B.* 360: 2011-2020.

Medelson, R., W. D. Nordhaus and D.Shaw (1994). The impact of global warming on Agriculture: A Ricardian analyses. *Amer. Econ. Rev.* 84:753-771.

Menzel, A and N. Estrella (2006). In : Finger prints of climate change- adapted behaviour and shifting species pattern (Walther, G.R., Burga, C.A. and Edwards, P.J.), Kluwer Academic, New York, pp.123-137.

Pandey, D.N. (2002). Global climate change and carbon management in multifunctional forests. *Curr. Sci.* 83:593-602.

Paul, B. and H. Rashid (1993). Flood damage to rice crop in Bangladesh. *The Geographical Review* 83: 151-159.

Parmesan, C. and G.Yuhe (2003). A globally coherent fingerprint of climate change impacts across natural systems. *Nature* 421:37-47.

Parry, M.L., J.H. Porter and T.R. Carter (1990). Climate changes and its implications for agriculture. *Outlook on Agriculture* 19:9-15.

Ruttan, V.W. (1999). The transition to agriculture sustainability. *Proc. Natl. Acad. Sci. USA* 96:5960-5967.

Sathaye, J.P.R. Shukla and N.H. Ravindranath (2006). Climate change, sustainable development and India: Global and National concerns. *Curr. Sci.* 90(3): 314-325.

Sinha, S.K. (1997). Global change scenario: current and future with reference to land cover change and sustainable agriculture - south and south-east Asian context. *Curr. Sci.* 72: 846-854.

Strokes, E. (2006). New diseases endanger Florida's already suffering citrus trees. *Science* 312:523-524.

Walther, E.R., E.Post, P. Convey, A. Menzel, C. Beebee, T.J.C. Parmesan, J.M. Fromentine, C. H. Guldberg, and F.Bairlein (2002) Ecological response to recent climate change. *Nature* 416: 389-395.

Wuethrich, B. (2000). How climate change alters rhythms of the wild. *Science* 287:793-795.

Zeigler, R.S. and D.W. Puckridge (1995). Improving sustainable productivity in rice-based rainfed lowland systems of South and Southeast Asia. *Geojournal* 35: 307-324.

Chapter-18

Traditional Approaches for Conservation and Utilization of Natural Resources in Chamba and Kangra Districts of Himachal Pradesh

VIPAN GULERIA AND ATUL GUPTA
Regional Horticulture Research Station Jachh, Kangra, Himachal Pradesh-176 201

ABSTRACT

Chamba and Kangra districts of Himachal Pradesh are abode to vast natural diversity. The people of the area had developed varied traditionl approaches to live with the nature so that natural resources can be protected for their sustained use. The present paper enumerates some of the ecofriendly life styles, agroforestry systems and water use practices being adopted by the farmers of the region.

Introduction

Himachal Pradesh is spread over an area 55,673 km^2. The land use pattern of the state is given in table 1. The state is land locked and is bordered by Jammu and Kashmir on north, Punjab on the south-west, Haryana on south, Uttarakhand on the south east and by Tibet on the east. Himachal is a mountainous region, rich in its natural resources. Most of the people are agriculture dependent. Over 93% of the population depends directly on agriculture. The physical features and climatic conditions of the region are not favourable for production of food grains. Cultivation of food grains on slopes of hills is neither economic nor beneficial. Hence the only option is to cultivate cash-crops.

Table 1 : Land Use Pattern of Himachal

Contents	Area (ha)
Total geographical area	
By professional survey	55,67,300
Forests	10,86,548
Barran and uncultivable lands	7,75,896
Land put to non-agriculture use	2,24,791
Permanent pastures and other grazing lands	13,40,222
Land under Misc. trees crops not included in area sown	97,527
Cultivable waste lands	1,03,942
Current fallows	22,706
Other fallow lands	55,410
Net area sown	5,60,154
Area sown more than once	4,26,184
Total cropped area	9,86,338

The main cereals that are grown in the state include wheat, maize, rice and barley. Although the state is deficit in production of food grains but it has been successful in other production spheres viz. seed-potato, ginger, vegetables, vegetable seed, mushrooms, chicory seeds, hops etc. Fruit cultivation is another field which is economic as well as favourable according to the agro climatic condition. The state is the major producer of apples in the country and hence named as the "Apple State of India".

The main sources of irrigation are *Kuhls* (small water channels) available from seasonal springs. Lift irrigation is another source of irrigation in the state. Efforts are being made by the government to provide irrigation facilities in most of the areas.

Elevation ranges from 450 meters to 6500 meters above sea level. The mountainous region of the state has three divisions from south to north: The outer Himalaya (Shivaliks), The lesser Himalaya (central zone) and the Great Himalaya (northern zone). The Shivalik range consists of lower hills (elevation – 600 m above sea level). The hills of the region are composed of highly unconsolidated deposits which results to a high rate of erosion. There is an increase notified in the elevation from west to east and from south to north. The lesser Himalayas are spotted by a gradual elevation towards the Dhauladhar and the Pir Panjal ranges. The rise is more rapid in the Shimla hills, to the south of which is the high peak of church – Chandni (3647 m). On north of river Sutlej, the rise is steady. The Kangra valley is a longitudinal trough which is at the foot of the Dhauladhar range. Dhauladhar which means the White Peak has a mean elevation of nearly about 4550 meters. It has a rapid rise of 3600 m above the Kangra valley. The largest of the lesser Himalayan ranges that is the Pir Panjal, branches off from the Greater Himalayan range near the Alpine zone, Kalpa in Himachal Pradesh. The Great Himalayan range (5000 to 6000 meters) runs along the eastern boundary and is slashed across by the Sutlej. Some of the famous passes in this range are Kangla (5248 m), Bara Lacha (4512 m), Parang (5548 m) and Pin Parbati (4802 m). The Zaskar range, the eastern most range separates Kinnaur and Spiti from Tibet. It has peaks rising over an elevation up to 6500 m. Some of the well known peaks are Shilla (7026 m) and Riwo Phargyul (6791 m), which also are the highest among the peaks in this range. There are many glaciers over the Zaskar and the Great Himalayan ranges. Himachal is also well known for its rich flora. In total Himachal have 49 cities and towns. The smallest town is Naina Devi and the largest is Shimla with a population of about 6,077,248. Urban population is only 7.5% of the total state population. Most of the state population still resides in rural areas. Himachal Pradesh is blessed with a huge area under green forests. A total area under forest is about 21,325 sq. km i.e. about 38.3 % of the total geographical area of the state (Table 1). Earlier forests were considered as the main source of income of the state and were exploited extensively. However, now the stress has shifted from exploitation to conservation. Still forests fulfil nearly 80 per cent needs of the local people. The forests of Himachal Pradesh are rich in vascular flora, which forms the conspicuous vegetation cover. Out of total 45,000 species of plants found in the country as many as 3,295 species (7.32%) are found in the State. More than 95% of species are endemic to Himachal and characteristic of Western Himalayan flora, while about 5% (150 species) are exotic introduced over the last 150 years. Himachal Pradesh is poor in mineral deposits. The state accounts for only 0.2% of the output of minerals in the country. There are some deposits of limestone (light grade), quartzite, brass, sand, gold, pyrites, copper, rock salt, slates, natural oil, gas and mica, barytes and gypsum which are commercially valuable.

Methodology

The survey and interaction regarding the unique ecofriendly approaches for routine life styles were first enumerated based on the experience. The areas of study were chosen after preliminary survey of the areas already marked with unique ecofriendly life styles. Detailed survey regarding the activities were done and the people following that particular activity were asked question regarding the procedure involved and the possible benefits they observe and then the artisans and other people who are involved in the formation of the activity were also asked to explain the whole process. The information was then cross-checked with the earlier available information.

Ecofriendly Life Styles of Gaddies and Gujjars of Chamba

House of Churah valley and Bharmour valley of District Chamba is the abode of the Gaddes and Gujjars, who are semi-agricultural tribe. Gaddis own and rare large flocks of goats and sheep. They spend most part of their life in transporting the animals from temperate to hot region and from hot region to temperate region in search of palatable grasses and it is their main source of income. Summers are spent in the higher mountains of Chamba, Pangi and Lahul and the winters in bordering plains. They bear the typical dress comprising *topa, chola* and *dora.* They are seen carrying hukah (smoking pipe) and spinning wheels in their hands and khalru containing cereals and other essential articles on their back. The Gaddis are simple and honest lot.

The life style of the Gujjars is almost similar to the gaddies except that they rear buffaloes and keep moving in search of grazing pastures throughout the year. In summer they go to the *dhars* (peak of mountains) while in winter they come down to the plains. They also do agriculture along with rearing buffaloes. They build there temporary structures at the grazing places called kothas. The Gujjars trace their origin to Gujarat. Certain unforeseen reasons forced them to emigrate from Gujarat. The Gujjars are simple, sturdy and hardworking people. They wear typical beard and dress. They wear on their heads a special type of turban (*safa*) made of muslin cloth mostly in white colour. As upper garment they wear a loose shirt (*kameez*) of blue or black cotton fabric.

The traditional houses in these areas are the excellent example of ecofriendly architect. The house is made up of stone and mud. Rooftop base of the house is made up of deodar, kail, fir, spruce and oak wood called "*shateers*". These are crisscrossed and above it the twigs of a fern, found only in temperate region of Chamba, are used to thatch. The fern used for the purpose has special quality as it does not deteriorate for many years. A layer of nearly 20 cm is laid above it, which is made up of the twigs of Texus, fir and spruce. The structure so formed is covered by soil and pine needle mixed up by the villagers and trampled for days so that it becomes stable and strong enough to bear the pressure of rain drops. The beauty of the whole process is that the whole village is engaged to construct the house without charging any money. This is the excellent example of social harmony and brother hood. Since these structures are made up of wood and soil therefore are bad conductor of heat and cold and make the house insulated from climatic factors and are also ecofriendly. Architcturally these structures are excellent as there is one small entrance on street side followed by a small room and one big kitchen in the centre of the house where there is "*chulah*" (fire place), with its both ends open (Fig. 1). There is ventilation just above it for the exhaustion of the smoke. It is made up in such a way that it may work even if there is heavy snow fall in these areas. Women keep their children on the cot while they work such as weaving, cooking, washing utensils etc. during winters when the whole area is snow covered. The houses of the area with their roofing made of mud is shown in photo 1.

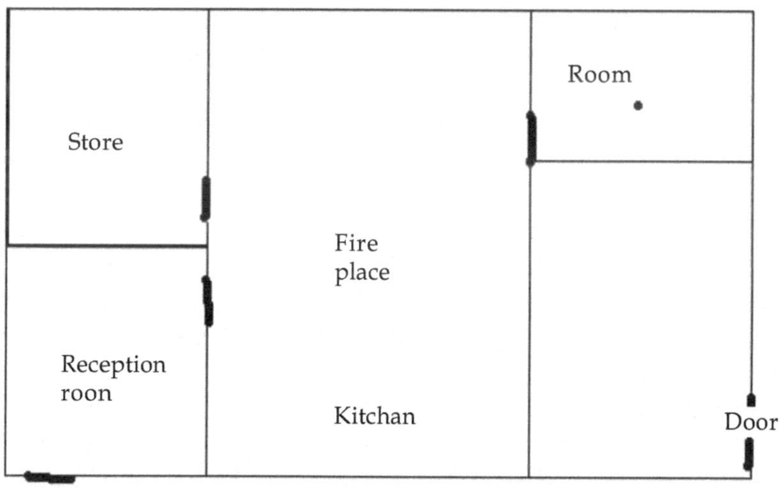

Fig. 1 : Outlay of the traditional house in rural Area Chamba

Photo 1 : Roofing pattern of the mud houses

Existing Agroforestry Systems of Chamba and Kangra

The survey of Kangra and Chamba district clearly mark the existence of four major agroforestry systems. The most common agroforestry system of the area is crop-based agroforestry systems where the major component is the cereal crops viz wheat+ maize as annual crops and *Grewia optiva, Toona ciliata, Delbergia sissoo, Abizzia lebbek* and *A. chinensis* are grown as tree crops on the bunds of the terraced fields.The survey of the area reveals that the most predominat species is the *Grewia optiva followed by Toona ciliata* >*D. sissoo*>*A. lebbeck*>*A. Chinensis* . The second most common existing system are horticulture-based agroforestry systems. The major component of these systems are the mango and citrus in low hill region, litchi in mid hill region and in high hill region apple, plum and peach are the major horticultural crops. The annual crops varies with the altitude, at lower elevation wheat, barseam and setaria are grown along with the horticultural crops. In mid hill regions wheat, rice and guinea grass are taken along with the horticultural crops. In high hill region the naturally growing local grasses and high yielding grasses such as fescue, festuca and timothy etc. are grown along with the horticultural crops. The unique crop combination is found in the high hill region of Chamba where there is special preference for the maize crop and it is grown along with the apple crops on the sloppy terraces. This is very much common in Tissa and Bharmour areas. The third agroforestry systems of the area are forage-based agroforestry systems. The fodder species such as wild olive, *Grewia optiva, A. chinensis* and *A. lebbek* are found growing through natural regeneration in the grass land of the region. The herbaceous crops are the naturally occurring grasses such as *Chrysopogon montanus, Heteropogon contortus, Apluda mutica* and *Thimeda arundenella* etc. The fourth agroforestry system prevalent in district Kangra is tea crop based agroforestry system where the major component is the tea crop and the woody component is the *Albizia chinensis. Albizia chinensis* is beneficial to the tea crop as it provide shade which increase the quality of product and it also provide fuel wood and timber.

Use of Wild Plants as Food of Tribal Areas of Chamba

"Ain" (*Girardiana diversifolia)* is a thorny shrub found in the sub temperate to moist temperate region of the district. It is very good food item and is part of the staple food of the tribal regions. The local preparation of it is called as "saag" .It is eaten with maize through out the year. The preparation also helps to fight against cold. Fern in its juvenile form is eaten in the form of dish. The juvenile form of the fern found in the sub temperate region of the district is called "linguru". It is cut into pieces to make pickle. Mushroom (*Genoderma* spp.) found on the dead tree trunk is used to make the tasty dish.

Ecofriendly Use of Water

Kuhl

Kuhl (canal) system is the only traditional irrigation system available in Himachal. The main feature of these *kulhs* is that almost all of them are driven by gravity. The water is tapped at one point of the stream by making temporary dam and small channel is made along the stream . Flow of water is maintained and this is taken to the fields in the valley. Almost 90 % of the total irrigated land area in Chamba district of Himachal Pradesh is covered by gravity irrigation. Many of these kulhs are almost hundred years old and are maintained by the local people at their own, but modern life style is creating problem in the management as people assume that it is the responsibility of the state government to maintain them. In almost all the villages of Churah and Bharmour there is '*Gharat*' (water mill) to make flour of wheat, maize and dalia. Many families make their both ends meet from these water run flour mills. Wood transportation is done though water which is another ecofriendly traditional method of wood transportation.

Khatri, bouwdis, Springs

Bamsan area of Hamirpur district in Himachals is tough area as far as natural resources are concerned because of concrete type of soil strata. There are scanty trees in the region. Water shortage is very common in summer. However, in the districts of Hamirpur and Sarkaghat subdivision of Mandi and Changer areas of Kangra district, the only source of drinking water is 'khatris' for their daily need.

Khatris are man made caves and can be found on roadsides in the Bamsan area of Hamirpur district. These khatris are made by digging the base of the hill and without disturbing the ground of the hill. The surface of the hill is covered with pastrues and small trees. The percolating water is harvested in them. Water of these sources is clean, cool and used by the local people for consumption. These khatris are the personal properties of some family and they are provided with iron gates and are kept under lock and key. Since there is no reliable source of water in the area, people are dependant on them.

There are two types of khatris found in Hamirpur, Mandi and Kangra districts. In one type, rain water is collected by harnessing water from roofs with the help of tins. This water is mainly used for washing clothes, utensils and for drinking by animals. In other type of khatris, only seeping water is collected and is used for drinking. These khatris are cleaned in winter and chlorinated. Bouwdies (shallow wells) dug out wells are another source of water harvesting in the state.

Soil Management Practices

Parts of Chamba districts had severe soil erosion (Photo 2). The best possible way out to check this erosion is by planting suitable trees and grasses and adopt sustainable agricultural practices.

In these areas soil is added in lower field from the bunds of upper field which in turn improves the nutrients and the soil fertility. The trees viz. *Grewia optiva, Quercus leucotrichophora, Olea europaea* (wild olive tree), *Toona ciliata, Bauhinia spp, Jatropha curcas, Albizia spp,*

Terminalia arjuna, Terminalia tomentosa , Terminalia belerica, Dodonea viscosa, Carisa carandus, Leucaena leucocephala , Dalbergia sissoo and *Acacia catechu* etc. are planted on the inner side of the lower terrace (Photo 3). *Grewia optiva* is the most common and highly planted tree around the bunds in agroforestry combinations; the species is highly regarded for good quality fodder feed in the winter when no other green fodder is available.

Photo 2 : Soil erosion a common phenomenon in Chamba District affected by heavy grazing and typical soil type

Photo 3 : Boundary protection from erosion by planting MPT's on the bunds

Bioengineering Approaches for Sustainable Watershed Management

The use of bio-engineering methods for soil erosion protection and slope stabilization has a long tradition. Old methods with rocks, plants, structures of timber have been used over the past centuries. Recently these old soil conservation and stabilization techniques have been rediscovered and improved. Bio-engineering methods have helped bridge the gap between classical engineering disciplines, land use management, landscape architecture and biological sciences. In the hilly regions of Chamba, Kangra and Hamirpur the use of stone, spurs etc. resulted in the heavy soil sedimentation in the subsequent years and are completely filled with the sediment. In some cases these have been broken by the force of the water and soil, however, the species such as *Ailanthus excelsa, Alnus nitida, Alnus nepalensis, Acacia catechu, Dalbergia sissoo and bamboos etc* had effectively stood against the force of water thus checking erosion (Photo 4).

Photo 4 : Bioengineering Structure a Success in Soil Erosion Control in Hills

There is traditional practice to keep surface maximum covered with grasses, shrubs etc., grazing is done in rotation and is allowed only during certain times. It is avoided during the flowering and seed setting stages of grasses. Fibrous rooted shrubs and grasses planted as hedges along the contour of the land slow down the runoff, weakens the erosive power of water and cause it to deposit its load of valuable soil behind the hedgerows. As a result the runoff proceeds gently down the slope where hedges have been planted at the correct vertical interval without erosive effect in the foot hills; erosive capacity of stream flow is also reduced by spurs of loose boulders. The ploughing of sloppy lands from bottom to top also helps in soil conservation as it checks the loosened soil strata falling from the upper side to the lower. The curved pattern is useful in maintaining infiltration rate of water which otherwise gets wasted with sudden runoff. The terraces are constructed across the slope i.e. along the contour. The terraces are supported by risers of suitable heights and width. The height of riser is again decided by the degree of slope. The risers are sometime made of loose boulders supported by grasses. The roots of grasses help in binding and keeping the boulders intact at a place. The roots of grasses help in drainage of excess water. With the traditional knowledge, farmers are keeping the risers toward inner slopes. In paddy growing areas the risers are erected to facilitate the pounding of water in the field. This type of bunding and terracing is continuing from centuries and terraces are still intact. The bunds are again used for growing palatable grasses which are used as fodder for livestock and trees are meant for fuel, fodder and fibre. The examples are beul, shisham, mango etc. In lower areas the bench terraces are known as "*khet*". Sometime on the risers contour hedge of grass like *khus*, local grasses are also established.

Chapter-19

Carbon Sequestration Potential of Production Forestry for Mitigating Impact of Climate Change

S.C. SHARMA*, A.S. DOGRA AND SANJEEV K. CHAUHAN*****

* Director, Zonal Research Station for Kandi Area, Ballowal Saunkhri, Distt. Nawanshahr, Punjab. ** Principal Chief Conservator of Forests (Retd.), Punjab State Forest Department, 17 Bays Building, Sector 17, Chandigarh – 160 016 *** Associate Prof. Forestry, Punjab Agril. University, Ludhiana- 141 004

ABSTRACT

The increasing level of green house gases in general CO_2 in particular has raised concerns about the vulnerability of human, animal and plant life. These concerns have prompted the Governments world over to take the appropriate measures to save them from perspective destruction. Forests though serve as source and sink of the CO_2, there is need to synthesize the impact of climate change on forests and the role of forests in global carbon cycle. Climate change needs to become part of decision making in forest management. Also, the agroforestry, a potential carbon sequestration activity is lagging behind, there is a growing interest in it in countries that have not only ratified Kyoto Protocol but also facing many other ecological problems on their private working land. The Clean Development Mechanism (CDM) offers benefits through sink credits and reduces the negative impact of climate change, therefore, necessary institutional arrangements should be put in place to use CDM projects effectively for generating resources through afforestation and reforestation projects for the well being of present as well as future generation. Forest/agroforestry scientists/managers need to realize the current degree of vulnerability and start thinking about their ability to cope with the impacts. Governments have initiated negotiations on policy actions to reduce the green house gases and addressed the vulnerabilities of the ecological, economic and social systems to climate change. It may be one of the least expensive ways for industries to off-set their carbon emission and tree growers to benefit through the carbon trade. Despite the fact that the certain aspects under Kyoto Protocol are still under clarification and negotiation, the forest plantations will play an important role in carbon sequestration and the implementation of Kyoto Protocol. Since the development of plantation is a slow process, the dividends will only be realized after a gap therefore the new plantation should be taken up without delay and the old one be managed as per their silvicultural rotation.

Introduction

During the past 100 years we humans, as a result of burning coal, oil and gas and clearing forests have greatly changed the chemical composition of the thin layer of lower atmosphere. These changes have far reaching consequences for the climate of the Earth, the ecosystems that are sustained by climate, our own health and economy. Although three primary gases - nitrogen, oxygen and argon make up 99.9 per cent by volume of Earth's atmosphere, it is the rare trace gases, the carbon dioxide, chloroflurocarbons and ozone that have the greatest effect on our climate. These trace gases are known as green house gases (GHGs). Of the main green house gases, CO_2 accounts for 63 per cent, methane 24 per cent, nitrous oxide 10 per cent and the remaining 3 per cent. The important feature of the GHGs is that they absorb certain infrared wave lengths and thus trap heat in the troposphere and stop it from escaping to space. Thus, it is important to direct efforts towards mitigation of CO_2, methane and nitrous oxide in particular.

Historically greenhouse gas concentrations in the Earth's atmosphere have undergone natural changes over time and those changes have been closely followed by changes in climate (Table 1). However, these changes were part of natural cycles and occurred over ten of thousands to millions of years. Recent human-induced changes, in atmospheric chemistry have occurred over decades (Ramanathan, 1988). This is referred to as "Climate change," which is defined as a change of climate that is attributed directly or indirectly to human activity that alters the composition of the global atmosphere and which is, in addition to natural climate variability, observed over comparable time periods. The global warming potential of carbon dioxide was first estimated by the Swedish chemist Svante Arrhenius. By 1938, some scientists concluded that human consumption of fossil fuels was already leading to significant increase in atmospheric CO_2 and global average temperature (Callender, 1938).

Table 1 : Greenhouse gases level since 1750

Greenhouse gas	Level	
	1998	1750
Carbon dioxide (CO_2)	365 ppm	87 ppm
Methane (CH_4)	1745 ppb	1045 ppb
Nitrous oxide (N_2O)	314 ppb	44 ppb
Tetrafluoromethane (CF_4)	80 ppt	40 ppt

Hexafluoroethane (C_2F_6) and Sulphur hexafluroride (SF_6) remained almost constant at 3 ppt and 4.2 ppt, respectively (Sharma, 2007).

Carbon dioxide is the largest single green house gas currently trapping about half of the total heat contributing to global warming. The average annual emissions of carbon were estimated to be 7.1 thousand million tons carbon per year and global carbon emissions from fossil fuel alone are estimated to be 5.5 thousand million tons per year and are likely to increase.

Recent research findings suggest that the global temperature is increasing and the year starting from December 1, 1997 to December 1, 1998 was the warmest year on record. Last decade of the last century was the warmest and the nine out of the ten years of the decade were the warmest in the recorded history (Stern, 2007). Although the global temperature is expected to increase by only about 1.4 to 5.8 °C by 2100, the impact of this on 'global climate change' is likely to be very serious. An another study in UK predicted the rise to be as high as 11°C. Global warming has the potential for significantly alter 35 per cent to the world's existing terrestrial habitats during this century. Some of the potential impacts of climate change have been evaluated by IPCC and discussed by Houghton *et al.* (2001) and Innes (2005).

- Catastrophic changes are likely expected in climate with added evaporation and precipitation, and vegetation movement towards poles by 200 to 300 Km. Each 1°C rise in temperature will displace the limits of tolerance of land species some 125km towards the poles, or 150m vertically on mountains (Bhadwal and Singh, 2002).

- For biodiversity, of the 2,70,000 known species of higher plants, 34,000 are endangered and in the tropics, ecosystem destruction may be so severe that some 60,000 species could be lost within next 25 years. About 5,200 species of animals are threatened with extinction. Today two of every three species are estimated to be on decline. Climate change poses a great threat to their survival than does the destruction of their natural habitat (Thomas *et al.,* 2004). The lower level of genetic diversity will be affected more adversely. Climate change may also lead to substantial changes in composition of forests (Dixon *et al.,* 2003).
- About 1964 million ha of land is faced with degradation, with 40 per cent of this is in Asia alone; 56 per cent of soil degradation is on account of soil erosion, which is as severe as 80 tonnes per hectare in certain areas.
- Adverse impacts on human health, sea level rise, agriculture production, forest fires, migration of population residing along sea coast, etc. are some other issues of major concern.

Analysis of the impact of climate change on forest productivity based on experimental research and modeling, suggest that forest productivity may increase under elevated carbon dioxide, but that the local conditions of moisture stress and nutrient availability will strongly temper any response (Joyce and Bridsey, 2000). In a review of woody plants, Ceulemans and Mousseau (1994) reported that mean biomass increment from elevated CO_2 was +38 per cent for coniferous trees and +63 per cent for deciduous trees. For coniferous trees, the range of responses was from +0 to +95 per cent. For deciduous trees, the range was from -47 per cent to +290 per cent. Summarizing studies not involving stress components, Saxe *et al.* (1998) reported larger average long-term biomass increment differences under elevated CO_2 for conifers of +130 per cent and smaller averages for deciduous trees (+49 per cent). However, Joyce and Nungesser (2000) concluded that the growth response to timber management is larger (25-39%) than the projected productivity response from changes in CO_2 or climate (8-29%). The outcome of enhanced carbon dioxide in the atmosphere is far from clear in the literature - there are complicated interactions between carbon dioxide and other site factors, which are still poorly understood. Carbon enrichment will have a greater effect on some species than on others. Oren *et al.* (2001) reported undetectable increment in loblolly pine due to increased atmospheric CO_2 at a nutritionally poor site. Large synergistic gains were detected with nutrient added sites, the gain being larger at the poor sites than at the moderate site. However, in 3-5 years, growth returned to previous level. Montagnini and Nair (2004) suggested that the carbon sequestration is not as straight forward and simple as it is often portrayed, therefore, the estimation of carbon sequestration should consider the limitations imposed by soil fertility/moisture as well as interactions with climatic factors.

Carbon Sequestration

There is a growing international scientific effort to develop ways to slow the addition of carbon dioxide to the atmosphere. One way to manage carbon is to use energy more efficiently to reduce our need for a major energy and carbon source, fossil fuel combustion. Another way is to increase our use of low-carbon and carbon-free fuels and technologies (nuclear power and renewable sources such as solar energy, wind power, and biomass fuels). The third and newest way to manage carbon is through carbon sequestration. Plants grow through the well-known process of photosynthesis. Plants absorb 3.67 units of CO_2 during photosynthesis to produce one unit of carbon that is stored in plant tissues.

Carbon sequestration especially sponging up carbon dioxide emissions with trees and other plants appears to be a simple, common sense approach to the climate crisis. It refers to the provision of long-term storage of carbon in the terrestrial biosphere, underground, or the oceans so that the buildup of carbon dioxide (the principal greenhouse gas) concentration in the atmosphere will reduce or slow down. The Intergovernmental Panel on Climate Change (IPCC) defines it as the "amount of carbon absorbed from the atmosphere by plants (equivalent to net photosynthesis), but which can then be lost in part to the atmosphere through subsequent decomposition of dead residues". It plays an important role in the global carbon cycle. The fact that we have temporarily thrown the global cycle out of balance suggests that we can also rebalance the system by improving vegetation management and natural sequestration of atmospheric carbon through photosynthesis. Obviously, the more trees and plants we have on earth, the more carbon sinks we have.

Forest Production and Productivity in India

The forest and tree cover (FTC) in India is estimated to be 778, 229 km^2 or 23.68 per cent (FSI, 2005) as against the objective of 33 per cent FTC enunciated in the National Forest Policy, 1988. The Planning Commission has advised the Ministry of Environment and Forests to achieve this target by 2012. However, going by the current rate of afforestation and reforestation in the country this appears highly unlikely. On the demand and supply side, Tables 2 and 3 depict estimates of wood balance upto 2020.

Table 2 : Projected supply of roundwood and fuelwood (million m^3)

Year 2000		Year 2010		Year 2020	
Round wood	Fuel wood	Round wood	Fuel wood	Round wood	Fuel wood
29	44	50	54	60	60

Table 3 : Demand for wood (million m^3)

Industry	Year 2000	Year 2010	Year 2015	Year 2020
Pulpwood based	8.76	21.92	34.67	45.86
Panel wood based	11.55	18.82	23.96	30.53
Durable wood based	37.69	54.26	64.37	76.61
Total	58.00	95.00	123.00	153.00

Source: Ministry of Environment and Forests, Government of India (2004).

A perusal of the above tables clearly brings out a huge gap between demand and supply. Another estimate places the demand and supply gap at 37-38 million M^3 in the case of industrial wood and a staggering 121 million tonnes in the case of fuel wood (Mahapatra *et al.*, 2005). While most of the industrial requirement is met by imports, the excess demand for fuel wood is obviously met from agricultural wastes and unrecorded and sustainable harvests from forests. Production forestry appears to be a promising in meeting the rising domestic demand for industrial and small timbers as also for removing large volumes of GHGs from the atmosphere.

The core objective of sustainable development is to balance social, economic and environmental activities and capital to improve current human welfare, while ensuring a sound foundation for future generations to maintain or improve their welfare (Solow, 1993). National Forest Policy, 1988 suggested that afforestation activities be taken up on degraded lands, wastelands as well as private lands by involving all the stake holders including local communities, NGO's and the investors. The key elements of the strategy would include multistake holder partnerships involving the Forest Department, local communities, NGO'S and the investors, with clearly defined obligations and entitlements for each partner to derive environmental, livelihood and financial benefits and rationalization of restrictions on cultivation of forest species outside notified forests to encourage social and farm forestry.

This is a laudable objective and must be operationalized as soon as possible. The policy should be to promote investments in technology based plantations on degraded forest lands and wastelands and private lands. Farm and agroforestry and agro-horticulture plantations on cultivated lands have the potential to increase wood production substantially in a very short period. Already social forestry contributes substantially to national wood supplies. There is a large potential for reforestation under JFM on non-forest areas. Ravindranath *et al.* (2003) have highlighted the eligibility of JFM under Clean Development Mechanism.

Carbon Sequestration in Agroforestry Systems

Agroforestry is a dynamic, ecologically sound, natural resource management practice that, through the integration of trees on farms and in the agricultural landscape, diversifies and sustains production for increased social, economic and environmental benefits. Approximately 1.2 billion people - 20% of the world's population - depend to a large extent on agroforestry products and services for their survival (Anon., 2000).

Incorporating trees into farming systems leads to greater prosperity at the farm level. Trees provide farmers with marketable products - such as lumber, building poles, firewood, animal fodder, fruits, medicines, etc. - all of which earn extra income. They improve soil fertility by fixing nitrogen from the air and recycling nutrients from the soil, thereby helping to increase crop yields and helping to ensure stability of future production. Trees on farms help hold moisture where it is needed, reduce soil erosion and keep valuable topsoil in place, reduce the intensity of downstream flooding and maintain watershed building materials. They serve as live fence in semi-arid regions, protecting vegetable and cereal gardens that would otherwise be overrun by livestock. Trees increase ecosystem biodiversity above and below ground and they can help ameliorate global climate change by sequestering carbon – in their live biomass as well as in the soil – that otherwise would be added to the earth's atmosphere.

Several studies have shown that the inclusion of trees in the agricultural landscapes often improves the productivity of systems while providing opportunities to create carbon sinks (Maikhuri *et al.*, 2000; Pandey, 2002; Albrecht and Kandji, 2003; Verchot *et al.*, 2005; Pandey, 2007). An average carbon storage by agroforestry land use system has been estimated to be around 9, 21, 50 and 63Mg C ha^{-1} in semi-arid, sub humid, humid and temperate regions, respectively (Schroeder, 1994). The amount of C sequestered largely depends on the agroforestry system/model put in places, the structure and function of which is to be a great extent determined by environmental and socio-economic factors. Carbon sequestration potential of afforestation/reforestation is specific to the species, rate and management involved, and it is, therefore, variable. Carbon sequestration potential for agroforestry practices is even more variable, depending on the planting density and production objective of the system. It has been estimated that globally approximately 38 Gt (1Gt = 1 billion tons) of carbon could be sequestered over next 50 years i.e., 30.6 Gt by afforestation/reforestation and 7Gt through the increased adoption of agroforestry practices. The prominent role of forestry and agroforestry systems in carbon sequestration has increased global interest in these land use options to stabilize green house gases emission. Throughout the world the area under agroforestry is of the order of 400 m ha. It is estimated that an additional 630 mha of current cropland and grasslands could be converted into agroforestry (IPCC, 2000). Agroforestry system especially agri-silvicultural/horti-silvicultural system can be carbon sinks and temporarily store carbon, while other systems (ruminant based) are purely green house gas emission systems.

Actually, the carbon storage in plant biomass is only feasible in the perennial agroforestry systems (perennial-crop combinations, agroforestry, windbreaks, hedgerow inter-cropping, etc.), which allow full time tree growth where the wood component represents an important part of the total biomass. However, a very large area of new forest is required to have a significant effect on the concentration of CO_2 in the atmosphere. Moreover, the carbon sequestration in agroforestry systems is a dynamic process. At the initial stage, the C and the N from the vegetation and the soil is lost, then there is quick accumulation during fast growth phase and at maturity or harvesting when the system returns back to agriculture or new plantation, some part of the C is released back though major portion is stored in the stem and the root system. The root system in general contributes to the soil carbon and the major share of stem wood carbon goes to the durable products and on an average 50 per cent of the stem wood carbon is converted to products like lumber and plywood that have a long life span. The short rotation species most preferred in the agroforestry interventions are not well suited to long-term locking of carbon. For example, the productivity of eucalyptus/poplar is more than shisham and kikar in Punjab but its carbon sink is less because of its shorter rotation and the relatively lower storage time of the wood products.

Productivity of India's forests is less than half the world average i.e., 0.7 $m^3ha^{-1}yr^{-1}$ as compared to the world average of 2.1 $m^3ha^{-1}yr^{-1}$. On the other hand, productivity of genetically superior clonal plantations of fast growing species is of the order of 20-50 m^3 ha^{-1} $annum^{-1}$ (Lal *et al.*, 2006). In northwest India because of good soil and availability of irrigation, yields are typically over 40 m^3 ha^{-1} $annum^{-1}$. If forest productivity is to be increased only improved seed and planting material should be used. In fact subsidized or free supply of seedlings indirectly encourages low productivity. In North-West India, productive clones of both poplar and eucalypts have been identified and are being propagated on a large scale. ITC clones of eucalypts are also being tested against Punjab selections. Punjab has identified superior clones of shisham (*Dalbergia sissoo*) and research to discover better clones is on. Seed orchards of important plantation species have also been established. The potential for growing teak in farm forestry systems is also high. Profitable agroforestry models are being developed through painstaking research. These systems are as profitable or perhaps more profitable, than agricultural crops alone. Current *et al.* (1995) reviewed 56 agroforestry practices in eight countries and found that a majority was profitable and in 40 per cent of cases, financial returns were at least 25 per cent higher than alternative farming practices. Zelek and Shively (2003) reported that agroforestry systems are lower-cost carbon sequestration alternatives to pure forest conservation. Ginoga *et al.* (2004) also reported encouraging economic potential of agroforestry in carbon sequestration.

Sustainably managed agroforestry systems have substantial economic and carbon sequestration potential. The principal barrier to agroforestry management appears to be socio-political rather than economic. Because management requires long term investment, good planning, selection of appropriate species, management technique, technical guidance, legal constraints and also the marketing structure need to be taken care-of. However, under productive agricultural zone, agroforestry as such may not be an option due to lack of species that can viably generate the products currently demanded by the market but the emerging carbon market may provide a new agroforestry option for landholders, through carbon sequestration. However, the prices of carbon would need to be higher enough to make growing trees on farmland a worthwhile investment than existing landuses (Wise and Cacho, 2005; Flugge and Abadi, 2006). Agroforestry systems/models with perennial crops may be important carbon sinks, while intensively managed agroforestry systems with annual crops may be more or less similar to conventional agriculture. More over the cost of carbon sequestration through agroforestry appear to be on much lower side than other CO_2 mitigation options. Economic analysis showed that costs of CO_2 mitigation through agroforestry could be easily offset by the monetary benefits from agricultural and tree products (Verchot *et al.*, 2005).

The integration of trees in agricultural landscape inevitably results in diversified and sustainable crop production, in addition to providing a wide range of environmental benefits. The Report on the task force for greening India proposes substantial area (28mha) for agroforestry interventions on irrigated as well as on un-irrigated ecosystems besides 15 m ha under JFM. Sathey and Ravindranath (1998) on an average estimated 25tC per ha average carbon sequestration potential in agroforestry in India. Singh and Lal (2000) have estimated that by bringing 40 m ha degraded land under forest with 5.5 tonnes/ha, the carbon mitigation of 3.32Gt would be possible in next 50 years with annual reduction of about 0.072 Gt carbon. Further, Lal and Singh (2000) emphasized that with the sustainable annual plantation area increase of 0.5 M ha, our forests are expected to act as a net C sink in future. In a study in Uttar Pradesh by Singh *et al.* (2000), approximately 20 million tonnes of carbon has been estimated to be sequestered by farm forestry plantations. Maikhuri *et al.* (2000) found the above ground biomass accumulation in a central Himalayan agroforestry systems to be 3.9 t ha^{-1} yr^{-1} compared to 1.1 t ha^{-1} yr^{-1} at the degraded forestland. Zhou and Wang (1997), Peichl *et al.* (2006) and Chauhan *et al.* (2007) reported greater potential of intercropping systems in reducing atmospheric carbon dioxide concentration compared to sole cropping systems. However, estimates of agroforestry potential still required better research information, most of the information available is on biomass only. The soil accounts for a major organic carbon pool, which needs to be quantified in tree based systems.

Carbon sequestration through agroforestry in Punjab plains and Shivaliks

Due to fast agricultural and industrial development in Punjab, the agro-ecosystem of state suffers from serious environmental degradation. Very high cropping intensity (180%) practiced on 84 per cent of state land area, of which 90 per cent is irrigated has resulted in deterioration of soil health, lowering of water table and environmental pollution. The State has only 6.3 per cent of the area under forests. In a policy paper on diversification in land management, it has emphasized that at least 15 per cent of the State's area should be brought under trees, and the agroforestry remains the only option. Poplar based agrisilviculture system is becoming very popular amongst the farmers of the Punjab in particular and northwestern region in general. Crops like wheat, berseem, oats, turmeric, colocasia, sugarcane, fodder, etc. are being grown by the farmers in block plantations of poplar.

The state of Forests in Haryana is quite similar to Punjab and to a large extent a similar situation also prevails in the reorganized states of Uttrakhand and Uttra Pradesh. Agroforestry systems are also being practised in the states of Haryana, lower areas of Himachal and terai area of Uttrakhand and U.P. With intensive management of poplar, presently farmers are earning Rs. 20,000 to Rs. 25,000 per acre per year from poplar based agroforestry systems against approximately Rs. 12,000 to 15,000 from rice-wheat rotation. In the near future, the area under poplar based agroforestry system is expected to increase. Farmers are also diversifying their plantation programme by incorporating eucalypts (*Eucalyptus* spp.), teak (*Tectona grandis*), dek (*Melia composita*), etc. in their agroforestry interventions.

The afforestation and reforestation activities though are eligible under clean development mechanism but the agroforestry plantation at farmer level does not qualify at present for the same. However, the agroforestry plantation meet all the criteria agreed under Marrakech accord for inclusion i.e., area (0.05-1 ha), canopy cover (10-30%) and tree height (2-5m), therefore possesses huge potential of carbon sequestration. Even the Indian threshold communicated to CDM Executive Board (Minimum area 0.05 ha; tree crown cover 30% and tree height 5 m) are also with in the acceptable limits under commercial agroforestry in general and poplar/eucalyptus based agroforestry in the North-Western states (Gera, 2007). Gera *et al.* (2006) reported 8.21lakh tons of carbon sequestration potential in 10,658 ha poplar and eucalypt plantations in Punjab. Singh (2005) demonstrated the usefulness of tree species in improving the carbon stock. The higher litter addition had positive influence on soil organic carbon accumulation (Negi *et al.*, 2006). Ravindranath *et al.* (2005) and Gera (2006) have reported likely C-benefits from agroforestry plantations. Taking into account the large area under degradation in Shivaliks, with less potential for agriculture due to rainfed and undulating condition and frazile ecosystem tree based system are more feasible. These systems will certainly help in sustainable production and also sequester carbon.

Commercial Forestry and Sustainable Development

The national objectives of increasing forest cover and forest productivity for sustainable development can be achieved by promoting integrated development of plantations, agroforestry and the wood based industries. This will not only lead us to self sufficiency but also create a surplus for export.

Ravindranath *et al.* (2001) have projected the sustainable forestry and commercial forestry scenarios with the aim of meeting biomass demands through plantation and conservation forestry. The mitigation potential projected per ha for the period 2000-2030 was found to be the lowest for short rotation forestry (25 Mg C ha^{-1}) and highest for forest protection (176 Mg C ha^{-1}). However, net present worth of short rotation forestry was the highest. The authors do not seem to have taken into account the vast potential of clonal plantations in carbon sequestration. Their estimates are way below the estimates for humid tropics (IPCC, 2000) and German plantations (Burschal *et al.*, 1993).

The carbon content of poplar and *Eucalyptus* trees in Punjab have been estimated on the basis of oven dry weight density. The procedure adopted by Chauhan *et al.* (2007) was followed to workout the carbon sequestration in poplar and eucalyptus. The figures are 0.47 and 0.45 times the respective oven dry weights (Table 4). The estimates are in line with the national figure of 0.4524 (Chhabra and Dhadwal, 2004). The carbon sequestration potential of poplar and clonal eucalyptus have been projected in Tables 5 and 6. The potential of long rotation species though higher, yet the potential of poplar and eucalyptus at short rotation is substantial (Fig. 1). At the same tree dimensions, eucalyptus can sequester comparatively more carbon than poplar because of higher wood specific gravity. But, it is important to mention that the locking of sequestered carbon in short rotation forestry is essential otherwise it would become a source than a sink of carbon.

Table 4 : Different types of wood densities of eucalypts (*Eucalyptus tereticornis*) and poplar (*Populus deltoides*)

Wood density type	Wood density (Kg m^{-3})	
	Eucalypts	**Poplar**
Green density	1174±156	843 ± 31
Air dried density (12 weeks)	804±146	481 ± 17
Oven dry density	696±141	413 ± 15

Table 5 : Amount of carbon sequestered (Kg) in Eucalyptus trees

Diameter at breast height (cm)	Tree height (m)				
	15	**17**	**20**	**22**	**24**
12	26.6	29.4	33.7	36.3	39.1
15	41.2	45.6	52.2	56.5	60.7
17	52.6	58.4	66.7	72.2	77.6
20	72.4	80.3	91.8	99.4	106.8
25	112.2	124.5	142.4	154.0	165.5
30	160.6	178.1	203.7	220.3	236.8

An additional amount approximately 20-25 per cent is sequestered and fixed in the soil through root biomass.

Table 6 : Amount of carbon sequestered (Kg) in poplar trees

Diameter at breast height (cm)	Tree height (m)				
	16	18	20	22	24
16	28.42	30.34	32.19	*	*
18	32.41	34.66	36.82	38.91	*
20	36.56	39.14	41.63	44.04	46.37
22	40.84	43.78	46.60	49.34	51.98
24	45.26	48.55	51.73	54.79	57.76

* Trees normally not available in these categories. An additional amount approximately 20-25 per cent is sequestered and fixed in the soil through root biomass.

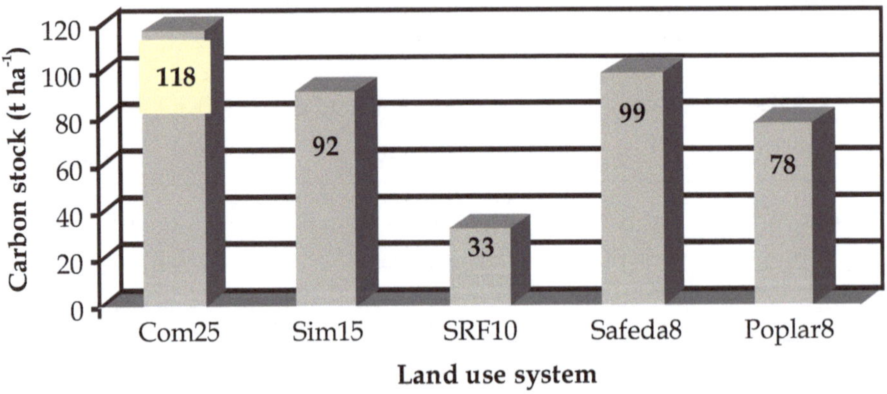

Fig. 1 : Comparison of carbon stock in different land use systems

Com25 = Complex agroforest duration 20-40 yrs
Sim 15 = Simple agroforest duration 15 yrs
SRF10 = Short rotation forestry 10 yrs
Safeda8 = *Eucalyptus* agroforest duration 8 yrs
Poplar8 = Poplar agroforest duration 8 yrs.

Source : IPCC (2000) and Punjab data.

The agroforestry and forestry holds potential in Shivaliks because it is a highly fragile ecosystem with soil loss up to 262 t ha^{-1} yr^{-1} in degraded Himachal Shivaliks. Out of 94.63 lakh ha degraded land, 26.94 lakh ha (assessed during 2003) was degraded due to soil erosion caused by water. Such erosion prone lands are not fit for agriculture production as it will further degrade them and produce less. Thus Shivalik lands can only be used sustainably by way of putting them under perennial cover either in form of agroforestry or under forest.

Conclusion

The review conducted by a distinguished international panel of scientists under the auspices of Intergovernmental Panel on Climate Change (IPCC) has determined that the balance of evidence suggests a discernible human influence on "global climate". The most efficient approach to slowing climate change is through market based policies. Countries that have joined the Kyoto protocol have, among others, agreed to promote and sustain carbon sinks.

Our contribution to the global CO_2 emission is approximately 4.07 per cent, which ranks at the fifth position. The developed countries are pressing hard to include developing countries like China, India, Brazil, etc. under the restricted list of Annex B industrialized countries, which will hamper the economic growth of the country but we have to share the concern and reduce the CO_2 emission. The industrialized countries alone cannot stabilize the global emission, the emission by developing countries would enhance the stabilization level inspite of zero emission by developed industrialized countries. Therefore, the climate friendly initiatives in the developing world itself are essential to halt the ever increasing level of GHGs.

Sponging carbon dioxide emissions with trees appears to be a simple and effective approach to the climate crisis. Technologies for increasing forest productivity are available. In order to exploit vast potential of carbon sequestration through agroforestry, innovative policies, based on rigorous research results have to be put in place. Government of India and State Governments should create an enabling environment by involving all the stakeholders - State Forest Departments, Corporate Sector, Rural communities and farmers to achieve a balance between social, environmental and economic objectives of sustainable development. Climate change as such cannot be looked in isolation, since the impacts are cross-sectoral.

The emphasis of climate change have much been focused on the reduction in concentration of GHGs in the atmosphere through reduction in emission or increased sinks. Instead, lot of focus is required in the adaptation to the changes that are occurring and predicted for future, regardless of mitigation strategies. Currently, no account is taken of climate change in silviculture. The forests are managed on the assumption that past performance is good indicator of future productivity. Forestry sector needs to pay much attention to climate change than it has till date for optimizing the productivity.

The afforestation and reforestation initiatives on a suitable scale accompanied by measures to halt deforestation could sequester carbon in sufficient amount to counter the increased level in the atmosphere, leading to global warming. The mitigation potential of forestry options are well understood but the plantation on the farms though sequester carbon directly in different bio-physical components but also reduce pressure on forest biomass for domestic and industrial requirements, thus maintaining the carbon pool intact in the living plant biomass, in the soil and in durable wood products. We need to strengthen growth and yield information and develop models for species of interest to communities. Production forests are capable of binding CO_2 but we have to be careful for the conservation of biodiversity. Unrestricted import of logs/timber under Open General License serves as a disincentive to agroforestry. At the same time there is a complete ban on export of wood. The ban may atleast be lifted on export of farm grown wood.

Agroforestry alone cannot solve the current climatic and socio-economic problems but can be one among the range of strategies with an added benefit of providing valuable products and food. Adequate understanding of agroforestry with regard to carbon sequestration will require more research. Policies that benefit climate change adaptation/mitigation, biodiversity and poverty reduction through afforestation and reforestation need the promotion at national level.

References

Albrecht, A. and S.T. Kandji (2003). Carbon sequestration in tropical agroforestry systems. *Agri. Ecosyst. and Environ.* 99: 15-27.

Anonymous (2000). Path to prosperity through agroforestry: ICRAF's corporate strategy 2001-2010. ICRAF Publication, Nairobi, Kenya. 43p.

Bhadwal, S. and R. Singh (2002). Carbon sequestration estimates for forestry options under different landuse scenarios in India. *Curr. Sci.* 83:1380-1386.

Burschel, P., E. Kursten, B.C. Larson and M. Weber. (1993). Present role of German forests and forestry in the national carbon budget and options to its increase. *Water, Air and Soil Pollution* 70: 325-340.

Callender, G.S. (1938). The artificial production of carbon dioxide and its influence on temperature. *Quarterly Journal of Royal Meteorological Society* 64: 223-240.

Ceulemans, R. and M. Mousseau (1994). Effects of elevated atmospheric CO_2 on woody plants. *New Phytologist* 127: 425-446.

Chauhan, S.K., V. Beri and S.C. Sharma (2007). Adhoc project report "Studies on carbon sequestration under different farm/agroforestry interventions". Dept. of Forestry & Natural Resources, PAU Ludhiana submitted to ICAR, New Delhi, 131p.

Chhabra, A. and V.K. Dhadwal (2004). Assessment of major pools and fluxes of carbon in Indian forests. *Climate Change* 64:341-360.

Current, D., E. Lutz and S. Scherr (1995). Costs, benefits and farmer adoption of agroforestry: Project experience in central America and the Carribbean. *Environmental paper No. 14*. The World Bank, Washington DC.

Dixon, R.K., J. Smith and S. Guill (2003). Life on the edge: Vulnerability and adaptation of African ecosystems to global climate change. *Mitigation and Adaptation Strategies for Global Change* 8: 93-113.

Flugge, F. and A. Abadi (2006). Farming carbon: an economic analysis of agroforestry for carbon sequestration and dryland salinity reduction in western Australia. *Agrof. Systems* 68: 181-192.

FSI (2005). State of Forest Report 2003. Forest Survey of India (Ministry of Environment and Forests), Dehradun. 134p.

Gera, M. (2007). Defining forest in Indian context as per CDM requirements. *Ind. For.* 133:1005-1013.

Gera, M. (2006). Role of farm forestry and agroforestry in carbon sequestration in India. *3rd Regional Conference on Scope of Production Forestry for Carbon Sequestration* organized by ATREE (Dec. 7-8, 2006), FSI Dehradun.

Gera, M., G. Mohan, N.S. Bisht and N. Gera (2006). Carbon sequestration potential under agroforestry in Roopnagar District of Punjab. *Ind. For.* 132: 543-555.

Ginoga, K., Y.C. Wulan and D. Djaenudi (2004). Potential of Indonesian small holder agroforestry in the CDM: a case study in upper Citanduy watershed area. *Working paper CC12*, ACIAR Project, ASEM 2002/066.

Houghton, J.T., Y. Ding, D.J. Griggs, M. Nogver, P.J. van der Linden, D. Xiaosu, K. Maskel and C.A. Johnson (2001). Climate change 2001: The scientific basis. Cambridge University Press, Cambridge, 881p.

Innes, J.L. (2005). The importance of climate change in considering the role of forests in the alleviation of poverty. 17th Commonwealth Forestry Conference (Feb. 28-March 5, 2005), Colombo, Srilanka, 16p.

IPCC (2000). Land use, Land-use Change, and Forestry. Watson, R.T., Noble, I.R., Bolin, Bert, Ravindranath, N.H., Verardo, D.T. and Dikken, D.J., (eds) A Special Report of the IPCC. Cambridge University Press.

Joyce, L.A. and R. Birdsey (2000). The impact of climate change on America's forests: A technical document supporting the 2000 USDA forest services RPA assessment. Gen. Tech. Rep. RMRS-GTR-59. Fort Collins US.133p.

Joyce, L.A. and M. Nungesser (2000). Ecosystem productivity and the impact of climate change. The impact of climate change on America's forests: A technical document supporting the 2000 USDA forest services RPA assessment, Joyce,L.A. and Birdsey, R. (eds.). Gen. Tech. Rep. RMRS-GTR-59. Fort Collins US, pp.45-68.

Lal, M. and R. Singh (2000). Carbon sequestration potential of Indian forests. *Environmental Monitoring and Assessment* 60: 315-327.

Lal, P., A.S. Dogra, S.C. Sharma and G.B.S. Chahal (2006). Evaluation of different clones of eucalyptus in Punjab. *Ind. For.* 132:1383-1390.

Maikhuri, R.K., R.L. Semwal, K.S. Rao, Singh,K. and Saxena, K.G. (2000). Growth and ecological impacts of traditional agroforestry tree species in Central Himalaya, India. *Agrof. Systems* 48: 257-271.

Mahapatra, S.P., N.K. Bhattacharjee, Soumitri Das and P. Upadhaya (2005). Scope of production forestry in enhancing carbon mitigation in India. Ashoka Trust for Research in Ecology and the Environment (ATREE), New Delhi.

Montagnini, F. and P.K.R. Nair (2004). Carbon sequestration : An underexploited environmental benefit of agroforestry systems. *Agrof. Systems* 61: 281-295.

Negi, S.S., S.B. Singh and M.K. Gupta (2006). Agroforestry: A sustainable land use for carbon sequestration. *Proceedings of the Workshop on Status and Potential of Agroforestry in North-Western Himalayas.* Jagdish Singh, V.P. Panwar, K.D. Singh and Joginder Singh (eds). Nov.14-16, 2006 at HFRI Shimla. pp.10-15.

Oren, R., D.S. Ellworth, K.H. Johnson, N. Philips, B. Ewers, C. Maler, K.V.R. Schaefer, H. McCarthy, H. Hendrey, S.G. McNutty and G.G. Katul (2001). Soil fertility limits carbon sequestration by forest ecosystems in a CO_2 enriched atmosphere. *Nature* 411: 469-471.

Pandey, D.N. (2002). Carbon sequestration in agroforestry systems. *Climate Policy* 2: 367-377.

Pandey,D.N. (2007). Multifunctional agroforestry systems in India. *Current Science* 92: 455-463.

Peichl, M., N.V. Thevathasan, A.M. Gordon, J. Huss and R.A. Abohassan (2006). Carbon sequestration potentials in temperate tree-based intercropping systems, southern Ontario, Canada. *Agrof. Systems* 66:243-257.

Ramanathan, V. (1988). The greenhouse theory of climate change: a test by an inadvertent global experiment. *Science* 240: 293-299.

Ravindranath, N.H., P. Sudha and S. Rao (2001). Forestry for sustainable production and carbon sequestration in India. *Mitigation and Adaptation Strategies for Global Change* 6:233-256.

Ravindranath, N.H., I.K. Murthy, P. Sudha, and C.A. Sahana (2003). Clean development mechanism and joint forest management programme in India. *Ind. For.* 129:815-825.

Ravindranath, N.H., I.K. Murthy, P. Sudha, V. Ramprasad, M.D.V. Nagendra, C.A. Sahana, K.G. Srivathsa and H. Khan (2005). Phase I report on *Methodological Issues in Forestry Mitigation Projects – A Case Study of Kolar District.* Centre for Sustainable Technologies and Centre for Ecological Sciences, Indian Institute of Science, Bangalore.

Sathey, J.A. and N.H. Ravindranath (1998). Climate change mitigation in the energy and forestry sectors of developing countries. *Annu. Rev. Energy Environ.* 23:387-437.

Saxe, H., D.S. Ellsworth and J. Heath (1998). Tree and forest functioning in an enriched CO_2 atmosphere. *New Phytologist.* 139: 395-436.

Schroeder, P. (1994). Carbon storage benefits of agroforestry system. *Agrof. Systems* 27: 89-97.

Sharma, A. (2007). If you think it's hot now. *Hindustan Times Sunday Magazine,* June 17, 2007, pp. 1-2.

Singh, G. (2005). Carbon sequestration under an agri-silvicultural system in the arid region. *Ind. For.* 131: 543-552.

Singh, R. and M. Lal (2000). Sustainable forestry in India for carbon mitigation. *Current Science* 78: 563-567.

Singh, T.P., Varalakshmi and S.K. Ahluwalia (2000). Carbon sequestration through farm forestry: case from India. *Ind. For.* 126: 1257-1264.

Solow, R. (1993). *An Almost Practical Step Toward Sustainability.* Resource Policy (Sept. 93). Resources for the Future. Washington, DC, USA, pp.162-172.

Stern, N. (2007). Freaky weather. *Hindustan Times* Jan.14, 2007, p.10.

Thomas, C.D., A.Cameron, R.E. Green, M.Bakkenes, W.Beaumont, Y.C. Collingham, B.F. Erasmus, M.F. Siqueira de, A.Grainger, L. Hannah, L. Huges, B. Huntley, A.S. van Jaarsveld, G.F. Midgley, L. Miles, M.A. Orlega-Huerta, A.T.Peterson, O.L. Phillips and S.E. Williams (2004). Extinction risk from climate change. *Nature* 427: 145-148.

Verchot, L.V., J. Mackensen, S.Kandji, M. Van Noordwijk, T. Tomich, C.Ong, A.Albrecht, C.Bantilan, K.V.Anupama and C. Palm (2005). Opportunities for linking adoption and mitigation in agroforestry systems. *Tropical forests and adoption to climate change in search of synergies.* Robledo, C., Kanninen, M. and Pedroni, L. (Eds.) CIFOR Bongor Indinesia, pp 103-121.

Wise, R. and O. Cacho (2005). A bioeconomic analysis of carbon sequestration in farm forestry : a simulation study of *Gliricidia sepium. Agrof. Systems* 64: 237-250.

Zelek, C.A. and G.E. Shively (2003). Measuring the opportunity cost of carbon sequestration in tropical agriculture. *Land Economics* 79:282-298.

Zhou, L. and H. Wang (1997). A simulation study on CO_2 assimilation and crop growth in agroforest ecosystems in the East China Plain. *J. Environ. Sci.* 9: 463-467.

Chapter-20

Rehabilitation of Degraded Lands through Silvipastoral Systems – A Balanced Approach for Alternative Land Use with Special Reference to N-W Himalayas

CHARAN SINGH AND O.P. CHATURVEDI
Central Soil and Water Conservation Research and Training Institute,
218, Kaulagarh Road, Dehradun-248 195

ABSTRACT

North-Western Himalayas have a large area which are degraded and bouldry. Such areas are not fit for crop production. These areas are generally kept unutilized and hence are invaded by uneconomic and invasive species. These lands can be brought under production though establishment of silvipastoral system which has a potential of fodder, fuel and timber production besides conserving natural resources. The paper discusses the techniques of establishment, management and production potential of silvipasture systems in north-western Himalayas.

Introduction

India has about 18% of the human and 15% of the livestock population of the world that needs to be supported from only 2% of the earth's land area. The per capita land availability has declined from 0.89 ha in 1951 to 0.37 ha in 1991 and is likely to be 0.20 ha by 2035 (Singh, 2005). It is estimated that demand for green and dry fodder would be short by 64 and 24% by 2020, while it was short by

62 and 22% in 2005. The current sustainable production of firewood is 17 million tons from forest areas and 98 million tons from non-forest areas, leaving a net deficit of 86 million tons which is removed illicitly from forest and scrub formations all over the country. Nearly 32.67% of India's geographical area is affected by various forms of soil erosion and land degradation. The area reported to be degraded in India varies widely, ranging from 63.9 million ha (Anonymous, 2000) to 187.70 million ha (Sehgal and Abrol, 1994), due to variations in approaches used to define degraded lands. The area under wastelands is estimated to occupy 63.85 million ha which is 20% of the country's geographical area. In the foothills of the Western Himalayas, large areas are categorized as bouldery wastelands which have been formed due to deposition of huge amounts of coarse material that has been brought down by seasonal torrents from the higher reaches. Deforestation, faulty landuse practices and road construction in the upper reaches lead to soil displacement which moves down the slopes and is deposited in the foothills in the form of gravels, with coarser fractions spreading over the original soil layers. Recent surveys indicate that the area affected by these deposits is about 2.73 million ha in the Himalayan foothills out of which 0.3 million ha occur in Doon Valley alone (Arora and Vishwanatham, 1995). These lands are unsuitable for cultivation of field crops and are invaded by uneconomic vegetation which needs immediate attention for rehabilitation to prevent further extension. Utilization of these degraded areas and wastelands in developing countries through silvipastoral systems is a widely recommended practice for meeting demands of firewood and fodder (Upadhyay and Pathak, 1994). Trees and grasses together simulate the multi-storeyed physiognomy of natural forests and help in the recovery of degraded habitats through nutrient cycling and enrichment (Nair, 1983).

The successful establishment of trees and grasses together depend on balanced access to solar energy, water and nutrients. Establishment and growth of herbaceous vegetation underneath trees is directly affected by tree density (Walker *et al.*, 1986), shade tolerance (Wolters *et al.*, 1982), light interception (Wilson, 1996) and moisture and nutrient availability (Raison *et al.*, 1992). Growth and herbage yield from the understorey component begin to be affected only when tree crowns are fully developed (Mathew *et al.*, 1992). In several cases, like water stressed sites, it has been observed that herbaceous vegetation affects tree growth due to competition for water and nutrients (Balocchi and

Phillips, 1997). Tree canopy management in a silvipastoral system is thus necessary to maximize system productivity, by lopping, pollarding, coppicing or thinning. Crown pruning is known to modify rooting patterns and processes occurring below ground (Jones *et al.*, 1998) in trees and probably enrich soil by greater fine root turnover (Jeschke and Hartung, 2000). Management of tree canopy by lopping and pollarding improves productivity of ground vegetation by enhancing light penetration (Deb-Roy, 1994). The impact of tree canopy management on grass productivity has been reported earlier from the north west Himalayas by Vishwanatham *et al.*, (1999), from semi-arid central India by Pathak and Roy (1987) and from the arid zone of Rajasthan by Shankarnarayan and Singh (1990). Perennial vegetation such as multipurpose trees (*Grewia, Bauhinia, Albizia, Robinia,* etc.) and hardy grasses (Napier, Gorda, *Panicum, Eulaliopsis,* etc.) are not only the alternative for economic utilization of these degraded lands but also protects them by considerably reducing runoff and soil loss.

Silvipasture is the most promising alternate land use system which integrates multipurpose trees, shrubs, legumes and grasses, mostly on non-arable, degraded and marginal lands for optimising land productivity. However, in arable land, boundary plantations, hedges and live fences, perennial crops with pasture, fodder banks, trees on cropland, hedgerow intercropping, shelter belts and windbreaks are the other form of silvopastoral practices. Silvipastoral practices help in conservation of natural resources and provide forage, timber and fuel wood on a sustainable basis. Small and marginal farmers who are engaged in raising livestock can benefit a lot by adopting this system.

Fodder Production from Marginal Lands

Marginal and sub-marginal lands are usually uneconomical for crop cultivation and are often under immense grazing pressure by livestock. As a result the productivity from natural grasslands is generally low. The research results of IGFRI, Jhansi and CSWCRTI, Dehra Dun indicate the possibility of obtaining forage productivity from lands under silvipastoral systems of the order 8 to 11 t ha^{-1} which includes forages from understorey grasses and also loppings from the fodder trees. The trees have also an additional advantage of providing green fodder during the extreme dry season. The system of integrating tree species along with arable crops/range grasses not only to serve purpose of conservation but also improve the productivity of the site.

Choice of woody species which form upper layer is very important as the overall sustainability of the system depends on it.

Fodder Production from Problem Soils

Nearly 50 per cent of land area in the country is in various stages of degradation. A large chunk of land under problem soils such as ravines, deserts, saline, alkaline and acidity, swampy and flood and cyclone prone conditions, waste areas along the canal and river banks, in watershed areas of the dams, areas along the railway line which all can form the reserves of forage production.

A number of crops have been identified for their suitability to be grown under saline and sodic conditions but it has been observed that the crops like berseem, teosinte, sesbania, trigonella, sainji can tolerate some degree of salinity and sodicity but karnal grass is very suitable under saline-alkali conditions. Some crops like teosinte and hybrid Napier can tolerate to some degree of water logging at latter growth stages. Similarly, para grass could be grown under marshy habitat.

Viable silvipastoral systems have been designed for better utilisation of alkaline soils. Abandoned alkali lands can be improved by growing *Prosopis juliflora* with *Leptochloa fusca* (Karnal grass). After 52 months with this combination, soil pH decreased from 10.3 to 9.4 and electric conductivity from 2.20 to 0.42 d S m^{-1}, while organic carbon in the top soil increased from 0.18 to 0.43%. These two species improved the soil to such an extent that it was possible to plough under the Karnal grass after 4 years and grow less tolerant but more palatable fodder species like *Medicago* and *Trifolium alexandrium* under *Prosopis* trees.

Reclamation of alkali soils by addition of organic manure, gypsum followed by crop of Sesbania (*Dhaincha*) and molasses have been carried out. The following filling mixtures for filling the planting holes in sodic soils have been suggested;

a) Shallow pit 3 kg gypsum + 3 kg FYM + 3 kg rice husk

b) Deep pit 5 kg gypsum + 5 kg FYM + 5 kg rice husk

c) Deep pit for fruit trees 10 kg gypsum + 10 kg FYM + 5 kg rice husk

Unlike alkali soils, amendments of gypsum or pyrites are not required in saline soils. Instead leaching with good quality water is done to wash out the salts from the rooting zones. Sub-surface planting (30-45 cm below) and ridge-trench planting method helps to keep the salts away from the rooting zones.

Development of Non-Arable Land for Productive Use

Cultivable wastelands are actually those lands where the productivity is less than its optimum and can be reclaimed and put to use for producing multiple benefits that shall benefit rural communities. Technologies for restoration of these lands exist and if properly implemented, wastelands can be restored to half their productive potential. Besides environmental stability, the restoration of wastelands provides employment opportunities and creates alternate sources of income for farmers.

While it is impossible to recreate the complex structure of a natural forest, properly planned mixed species plantations can serve as 'foster' ecosystems. Plantations that are raised, protected and managed efficiently can mimic natural forests to a large extent and perform the same functions. Thus trees and grasses can be managed to provide fuelwood, fodder, fruits and non-wood forest produce in wastelands of different types.

(A) If the objective is to improve degraded lands and thereby reduce runoff and erosion, then it is necessary to built and /or improve forest areas. This can be done by protection from grazing, fire; planting with suitable multi-purpose tree species; supplementing with contour staggered trenches, furrows etc.

(B) If the objective is to maintain good watershed conditions, while expecting timber from forest and grazing, then all activities have to be done in a controlled manner. Only 30-40% of the basal area (of trees) should be harvested and if soil losses are expected to be high than only 25-30% should be removed. Grazing by animals is to be controlled and pockets of natural regeneration are to be strictly protected. Trees have to be harvested by following the selection system.

The selection of species for plantation on degraded lands depends upon the choice of ultimate end user. For example, trees planted for fuel wood should be fast growing, capable of producing coppice shoots rapidly, have high calorific value wood and easy to harvest and transport. Trees planted for fodder should be fast growing, nitrogen fixing, rich in crude protein, produce nutritious non-toxic leaves and coppice easily. Species planted for timber production should be tall with straight stem and have strong fine grained wood with good machining ability.

Establishment of Silvipasture Systems

Silvipasture system establishment particularly in Class V, VI and VII lands, which are not suitable for crop production can be accomplished through pasture establishment, range improvement and grassland improvement measures.

A. Pasture Establishment

Silvipastoral systems can be raised on marginal lands with suitable fodder grasses like *Dicanthium annulatum, Leptochloa fusca, Brachiaria mutica, Panicum antidotale, Sehima nervosum, Chloris gayana, Cenchrus ciliaris.* Trees are planted at wide spacing of 5x5 m or more depending up on the area available. In the initial years, grasses are harvested manually and later on when the trees have attained sufficient height, animals are permitted to graze. Trees which can be planted on marginal lands for silvi-pastoral system are: *Acacia nilotica, A. catechu, Terminalia alata, Madhuca latifolia, Anogeissus latifolia, Adina cordifolia, Hardwickia binata,* etc. To improve grass yields and fodder quality of pastures, sowing/planting of legumes like *Stylosanthes hamata, Medicago* and *Trifolium alexandrium* can be done. Silvipasture systems involving *Dalbergia sissoo + Chrysopogon fulvus; Acacia catechu + Eulaliopsis binata; Bauhinia purpurea* or *Albizia lebbek* with *Eulaliopsis binata* were found promising in the utilization of degraded lands for fuel wood and fodder production.

B. Range improvement

In this kind of silvipasture system, trees are grown at wide space (8x8 m; 10x10 m or 12x12 m) in the rangeland (grassland) and controlled grazing by the livestock is allowed. The optimum productivity of the range is maintained by appropriate grazing systems with the required livestock and periodic improvement measures.

C. Grassland Improvement

1. Before taking up improvement, it is necessary that protection is provided to the area, which should not be a very large area in the first phase (4-5 ha), with the area increasing slowly in phases. Protection, with the active involvement of farmers is desired to

first ensure recovery of the site and secondly to avoid social conflict. Areas close to a village, with shallow soils (< 30 - 45 cm) and moisture stress for a longer period which cannot be put to other use, can best be developed as grazing areas. Use of thorny bushes, *Agave*, staking with dried thorny branches or stone walls are the common and cheap methods for protecting such areas.

2. To improve seed germination or establishment of slips, soil and water conservation in these areas becomes necessary. Contour furrowing, contour trenching and contour bunding (on 5-8% slopes) have been found to be effective in moisture conservation and the consequent increase in forage production. Furrows can be made by tractor drawn implements, with the contours being 5–7 m apart, depending on slope features and rainfall pattern. Yield increases, ranging from 100–165% have been reported from the dry regions.

3. Reseeding of grasses, although economical, does not guarantee seed germination if the soil condition is not conducive. The best time to sow seeds is after the first heavy shower. The seed is mixed with sandy soil 3-4 times the seed volume and sown by seed drills, 50-75 cm apart, 3-5 cm deep. Grass species suitable for planting or seeding along with their production potential in silvipastoral land use systems are - *Brachiara mutica* (5-7 kg ha^{-1}), *Cenchrus ciliaris* (4-6 kg ha^{-1}), *Themeda triandra* (7-8 kg ha^{-1}), *Chrysopogon fulvus* (5-7 kg ha^{-1}), *Panicum maximum* (4-5 kg ha^{-1}), *Panicum antidotale* (4-5 kg ha^{-1}), *Apluda mutica* (5-7 kg ha^{-1}) and *Bromus* spp (5-6 kg ha^{-1}) *C. setigerus* (5-7 kg ha^{-1}), *Lasirus sindicus* (6-8 kg ha^{-1}), *Dicanthium annulatum* (2.5-3.5 kg ha^{-1}). A list of suitable grass and legumes for different agro-climatic situations is given in Table 1. Periodic weeding is carried out frequently in the first year and continued till the stand is well established.

4. Planting of rooted slips of suitable grass species is costly but a certain method of grassland improvement. Slips are planted 0.5–0.7 m apart in shallow holes and pressed firmly into the ground. The entire area is covered in this way, after the first heavy showers, when the soil layer is moist. Application of DAP @ 20 k ha^{-1} is beneficial for growth. In difficult areas, pelleted seeds are broad casted in the beginning of the rainy season. Pellets are prepared by mixing seeds with cow dung, clay, sand in a ratio of 1:1:3:1, using water to prepare round pellets of 0.5 cm diameter.

5. Legumes are known to fix nitrogen in the soil and being rich in crude protein, they improve the fodder value when mixed with poor quality or dry grass. A list of the important legumes for different agroclimatic regions are given in table 3. Legumes are usually mixed with grasses in a ratio of 1:4. Application of small doses of DAP are required to allow for nodulation to occur in poor soil, where much of the top soil has already been washed away.

Table 1 : Suitable grass and legume species for different sites

Region	Grass species	Legumes
North eastern Humid region	*Arundinella bengalensis, Imperata cylindrica, Pennisetum purpureum, P. polystachyon, Imperata cylindrica, Setaria sphacealata.*	*Medicago sativa, Gycine javanica, Stylosanthes hamata,Stylosanthes gracilis, Dolichos axillaris*
Northern sub-humid region	*Dicanthium annulatum, Chrysopogon fulvus, Cenchrus ciliaris, Pennisetum pedicillatum, Setaria sphacealata, Panicum maximum, Hybrid napier*	*Stylosanthes hamata, Dolichos lablab, Phaseolus atropurpurem*
Semi arid (<800 mm/year)	*Sehima nervosum, Cenchrus ciliaris Dicanthium annulatum, Chrysopogon fulvus Eragrostis curvula,Chloris gayana, Panicum antidotale, Heteropogon contortus Brachiara mutica,Isaelema laxum*	*Stylosanthes hamata, Phaseolus atropurpureum, Dolichos lablab, Dolichos axillaris*
Arid (<500 mm/year)	*Lasirus sindicus,Cenchrus setigerus, Panicum antidotale, Cenchrus ciliaris, Dicanthium annulatum*	
Saline - alkali (Salt affected)	*Leptochloa fusca* (karnal grass), *Chloris gayana* (rhodes grass), *Cynodon dactylon* (bermuda grass), *Brachiaria mutica* (para grass), *Panicum maximum* (Guinea grass), *Pennisetum purpureum* (Hybrid napier).	*Desmodium* sp. *Rhychosia minima*

Table 2 : Promising trees, grass and legumes for silvipastoral systems in few regions of India

Agroclimatic zones	Fodder trees	Grass and legumes	
Temperate, (J&K, Himachal Pradesh and NW Himalaya)	**Wet region** *Robinia pseudoacacia, Celtus austratis, Populus alba, Quercus semecarpifolia, Q. dilatala, Acer campbelli.* **Dry region** *Salixfragilis, P. ciliata, P. euphratica, Belula utilis, Acer candatum, Q. ilex*	Lucerne *(Medicago satira)*, Red clover *(Trifolium pratense)*, white clover *(T. repens)*, Cocksfoot *(Dactylis glomerala)*, Timothy *(Pheleum pratense)*	Singh, 1986
Kangra Valley, HP	*Erythrina suberosa Dalbergia sissoo, A. catechu, Albizia stipulata*	*Eulaliopsis binata, Chrysopogon fulvus, Bauhinia purpurea, Albizia lebbek, Napier bajra, Setaria anceps*	Vishwantham *et al.*, 1998
North-eastern Hill region	*Albizia lebbek, Bauhinia purpurea, Leucaena leucocephala. Salix* spp., *Populus ciliata, Castanopsis, Celtis australis, Sachima vallichii, Ficus* spp., *Bauhinia variegata*	*Panicum maximu, Setaria sphacealata, Pennicetum polystachyon, Stylosanthes guyanensis,* Bajra x Hybrid Napier, *Panicum maximum* Hamiu *Doctylis glorneruta, Lolium perenne*	Singh *et al.*, 1996
High rainfall, Doon Valley	*Leucaena leucocephala, Grewia optiva, Acacia sp., Bauhinia retusa, Salix tetrasperma.*	*Eulaliopsis binata, Chrysopogon fulvus*	Dhyani *et al.*, 1988
Saline-alkali soils	*Prosopis juliflora, Acacia nilotica, Terminalia qrjuna, Pongamia pinnata*	*Leptochloa fusca, Chloris gayane, Brachiana mutica, Sporobolus sp.*	Singh *et al.*, 1994

Source : Singh, 1986

Table 3 : Promising top feed species for silvipastoral use in the subtropical zone

Species	Common name	Rating	Crude protein (%)
Acacia nilotica	Babool	Good	14
Acacia catechu	Khair	Good	13-18
Anogeissus latifolia	Bakli	Good	7.5-11.5
Ailanthus excelsa	Maharukh	Excellent	20
Anogeissus pendula	Kardhai	Good	8-10
Albizzia lebbek	Siris	Good	17-26
Bauhinia variegata	Kachnar	Good	11-16
Cordia dichotoma	Lasora	Good	12.4-15
Ehretia laevis	Chamror	Medium	13.5
Ficus glomerata	Umar	Good	11.2-15.2
Ficus lacor	Pakar	Good	16
Kydia calycina	Safed dhaman	Excellent	11-13.6
Lannea coromandelica	Jhingan	Good	11.4-16
Prosopis cineraria	Khejri	Excellent	15.2
Syzygium cuminii	Jamun	Good	10.2

6. Once planting/seeding has been completed, it is necessary to provide adequate protection (illegal grazing, fire, harvesting etc.) to the area in the first year. In the 2nd year, depending on the establishment, a few manual harvests can be allowed. From the third year controlled grazing can be permitted based on the carrying capacity. Even then a few small patches need to be rested and allowed to seed, so that seed collection for further multiplication, of the desired species can be carried out. After 5 years, the area needs to be examined once again for any improvement if required, which is most easily seen from decline in yields.

Selective Silvipastoral Practices for Arable Lands

A. Boundary plantations of fodder species

In this practice fodder species such as *Acacia* spp., *Adina cordifolia, Ailanthus excelsa, Albizia lebbek, A. procera, Anogeissus latifolia, Celtis australis, Dalbergia sissoo, Derris indica, Ficus* spp., *Grewia optiva, Leucaena leucocephala, Melia azadirach, Sesbania grandiflora, S. sesban* are planted along field boundaries or other borders eg. along footpaths or irrigation channels as per the agroclimatic condition.

B. Trees on erosion control structures

Trees such as *Leucaena leucocephala, Melia azedarach, Sesbania grandiflora, S. sesban, Gliricidia, Calliandra, Grevillea robusta, etc.* are added to earth structures employed for soil and water conservation, such as terrace risers, bunds, ditches and grass strips. These trees help in stabilizing the structures and make productive use of the land which they occupy. Grasses such as napier, guinea, vetiver are also planted in between the trees.

C. Wind breaks and shelter belts

Planting of suitable MPTS wind breaks around the field boundaries is very important for crop production. MPTS are planted in lines across the direction of damaging winds, to reduce wind erosion and wind or frost damage to crops. They also provide shelter for animals. Some of the trees suitable for wind breaks/shelter belts are: *Acacia senegal, Aegle marmelos, Albizia lebbek, Bauhinia purpurea, Cordia myxa, Dalbergia sissoo, Holoptelia integrifolia, Kigelia pinnata, Lagerstroemia indica, Leucaena leucocephala, Madhuca indica, Pongamia glabra, Prosopis cineraria, Tamarindus indica, Syzygium cumini,* etc.

D. Hedgerow intercropping and contour hedgerows

This is also called alley cropping or alley farming. Hedges of *Leucaena leucocephala, Sesbania grandiflora, S. sesban, Gliricidia, Erythrina, Albizia or Cassia* are planted as more or less parallel rows with the plants closely spaced and crops/grasses are frown in the alleys between them. The rows are generally kept 4-10 m apart. The hedges are regularly pruned. Prunings may either be removed as fodder and fuel wood or retained on the soil. A contour hedgerow is a variation of this system, practiced on sloping lands and with the primary objective of soil and water conservation.

E. Fodder Banks

Trees are planted as blocks and managed for fodder production. This can make good productive use of areas of poorer soil on the farm.

Management of Trees under Silvipasture Systems

The management considerations are influenced by site conditions, inputs, demand for forest products, market prices, etc. Trees have the potential to provide the maximum returns (through biomass), only when they are managed effectively and efficiently. Some of the popular techniques adopted to manage trees are given here.

A. Tending

Tending is carried out for the benefit of a forest crop by creating best possible conditions of growth. Tending helps in producing high quality wood and maximizing returns per unit area, it is therefore, an important silvicultural operation. Tending operation include weeding, cleaning, thinning, pruning, climber cutting, etc.

Immediately after the planting is over, pits should be mended, and if climatic conditions permit, casualties are replaced before fourth week of July to enable the seedlings to get sufficient moisture from the rains that follow. Limited casualty replacement is desirable in second year and only if there is a big gap due to insufficient moisture or other climatic hazards. Simultaneously, first weeding and hoeing is carried out around the plants within 2-3 weeks after planting. *In situ* water harvesting using the tie ridge technique is desirable to combat the moisture stress. The second and third weeding should follow mulching with dry grass/cut weeds around the pits. This operation should be completed before withdrawal of monsoon.

It is necessary to carry out the two or three weeding and hoeing during first year and another two during second and third year. The sites with woody weeds and scrub growth may need additional one or two shrub cuttings or pruning to keep the plants free from their adverse effect. Use of herbicide and weedicide may be useful particularly for the woody plants.

B. Canopy Management

In arable lands as the farm holdings are small and the farmers' first consideration is to grow food and he cannot afford to sacrifice crop yield for tree growth. Therefore, canopy management besides selection of compatible shade tolerant crops (millets, rhizomatous crop) and trees with minimum shade *(Albizia,* etc.) is the important consideration. Canopy management options in relation to trees include pruning, lopping, pollarding and coppicing. Trees in wood lots may be coppiced to produce fuel wood, while trees in fodder lots or hedge rows intercropping systems are often lopped to produce fodder and mulch and where necessary to reduce shade on the adjacent crops. Trees in pastures or on boundaries are pollarded for the same purposes. In crop lands, on boundaries or in home gardens, trees are pruned to reduce shading, to increase crop productivity and provide fuel wood and fodder.

C. Pruning

Pruning is removal of live or dead branches or multiple leaders from standing trees for the improvement of the tree or its timber. Pruning is the principal method for managing the vegetative yield of trees. Pruning stimulates the outgrowth of dormant buds that are often present at old nodes over the whole shoot system. Pruning near ground level produces longer shoots than pruning higher up. Pruning reverses tree ageing and the new shoots are vigorous in growth. Pruning operations are commonly done in forestry practices for improvement of timber volume of trees.

D. Lopping

Lopping means cutting of branches of a tree. Incidentally the lopped trees produce new shoots which are annually or periodically lopped for various purposes. *Acacia, Anogeissus, Hardwickia binata, Melia azadirach, Moms spp., Terminalia tomentosa, Kydia calycina,* etc. are annually lopped for leaf fodder. The optimum cutting frequency to ensure survival and rapid recovery will depend on the species, environment and amount of foliage left.

E. Pollarding

Pollarding in forestry is an operation in which the stem of a tree is cut off at a height beyond the reach of browsing animals with the object of producing a crown of new shoots from buds below the cut. The flush of new shoots is cut down periodically so that the pollard may produce fresh shoots again.

F. Coppicing

Coppice is to cut broad-leaved trees close to the ground to produce sprouts and regrowth. In fact in forestry practices, coppice is a method of vegetative reproduction in which trees when cut from near the ground level, produce coppice shoots. However, its success depends on the factors such as species, age of tree, season of coppicing, height and method of cutting and rotation. Regular coppicing promotes the growth of leafy shoots which are accessible for harvesting or browsing, and by restricting the development of the boles and woody frames, coppicing increases the proportion of dry matter availability to shoots. However, all species do not coppice and even in the species that coppice, the power varies with species. Some examples of good coppicer tree species are, *Dalbergia sissoo, Acacia catechu, Shorea robusta, Tectona grandis* and species of *Albizia, Salix, Morus,* etc.

Rehabilitation of Degraded Lands for Mutliple Outputs

A number of MPTs viz. *Grewia optiva, Morus serrata, Celtis australis, Robinia pseudoacacia, Ulmus wallichiana, Quercus* spp., *Bauhinia variegate, Melia azedarach, Parkinsonia aculeata. Eucalyptus tereticornis,* Cassia *siamea, Acacia auriculiformis, Dalbergia sissoo, Albizia procera, Acacia nilotica, Thespesia populnea, Ailanthus excelsa, Derris indica* etc. have been found suitable for establishment of silvipastoral systems for rehabilitation of degraded lands. Trees with forage are preferred so that they can be lopped during lean periods and roughages supplemented with palatable and nutritious leaves for livestock feeding. *Hippophae rhamnoides,* an important indigenous multipurpose shrub of J & K, is used as fuel and fodder by the local inhabitants. In Spiti Valley of Himachal Pradesh 1 ha plantation of this shrub can meet the fuelwood needs of about 20 families (ICFRE, 1993).

In general, trees for fodder should be fast growing, nitrogen-fixing, a good coppicing ability and production of nutritious leaves. A list of the tree/shrubs/grasses suitable for the given area which can be used for fodder is presented in Table 2. These species can be grown on a wide variety of sites and lopping can be carried out when the trees have attained sufficient growth.

Trees planted for fodder as top feed along with suitable grass species should be widely spaced, so that moderate shade is available. At scattered locations some closely planted, shady trees are required to be planted so as to serve as resting spots for animals, along with a watering hole (for drinking and wallowing). Species with top feed value (Table 3) are preferred, so that some fodder is also available during extended dry periods. A list of species suitable for planting as trees in wastelands is provided in Table 4. Many of these species have other uses as well. It is however advisable to mix these with other species so that farmers are involved in their protection. The list is not comprehensive and neither is it necessary to plant all of them. The final choice of species should normally be obtained from the end-users. Participatory approaches and joint forest management plans should come into play in the very beginning of all development projects.

Table 4 : List of fast growing tree species suitable for planting as permanent vegetation

Name	Common name	Habit	Coppicing ability	Uses	Min. Rainfall (mm)
Acacia nilotica	Babool	SP	Medium	Fw/Fd/T	200
Acacia senegal	Gum acacia	SE	Good	Fw/Fd/Gum	200
Acacia tortilis	Israeli babool	SP	Good	Fw/Fd	100
Ailanthus excelsa	Maharukh	SP	Good	Fw/Fd	600
Adina cordifolia	Haldu	SP	Medium	Fw/Fd	500
Albizzia lebbek	Siris	SP	Medium	Fw/Fd	400
Azadirachta indica	Neem	SP	Medium	Med/Fw	400
Boswellia serrata	Salai	SP	Medium	Fw/Fd	400
Cassia fistula	Amaltas	SP	Good	Fw/Fd/M	500
Cassia siamea	Kassod	SP	Excellent	Fw	600
Dalbergia sissoo	Shisham	SP	Excellent	Fw/Tim	400
Dalbergia latifolia	Rosewood	SP	Good	Tim/Fw	600
Diospyros melanoxylon	Tendu	SP	Good	Bidi leaves/Fw	500
Erythrina suberosa	Madar	SP	Good	Fw/Fd/T	500
Gmelina arborea	Gamhar	SP	Good	Fw/Fd/T	500
Leucaena leucocephala	Subabool	SP	Excellent	Fd/ Ml/sT	400
Hardwickia binata	Anjan	SP	Medium	Fw/Fd/T	400
Holoptellia integrifolia	Kanju	SP	Medium	Fd/Fw/T	500
Madhuca indica	Mahua	SP	Good	Fd/Fw/Fr	500
Terminalia belerica	Bahera	SP	Medium	Fr/ Med.	500

(SP- spreading crown; SE – semi erect crown;
Fw- fuelwood, Fd – fodder; T- timber).

In the Shivalik foothills, agrihorticulture and hortipastoral systems have been developed for the highly eroded lands. Intercropping of guar, cowpea or pearl millet with peach; turmeric with papaya; *Chrysopogon fulvus* or Napier grass with *aonla* or *ber* has been identified as an economically-viable and eco-friendly system for rehabilitation of these lands. Aonla gave the highest yield of 86 kg fruit tree^{-1} in association with pigeonpea but the yield was reduced by 17 and 23% respectively in association with *Chrysopogon* and Napier grass. Silvipasture systems involving *Eulaliopsis binata, Saccharum munja* or *Vetivaria zizanoides* with *Acacia nilotica* were also found promising. Further, *Eucalyptus tereticornis* + *Eulaliopsis binata* has been identified as an economically viable and eco-friendly system for rehabilitation of these lands.

For highly degraded lands in semi-arid regions of central India, silvipastoral systems involving multi-purpose species are most suitable. *Albizia amara* based silvipastoral system produced 7.95 t ha^{-1} yr^{-1} fodder followed by *Acacia tortilis* (7.60 t ha^{-1} yr^{-1}). At Jhansi in a silvipastoral system consisting of *A. amara* + *L. leucocephala* + *D. cinerea* + *C. fulvus* + *Stylos* and natural grassland the total biomass yield of 12.62 t ha^{-1} yr^{-1} (4.55 t from pasture + 0.51 t from tree leaves and pods through pruning + 1.66 t from fuelwood through pruning + 5.90 t from small timber, fuelwood, leaf fodder through harvesting of trees) was recorded (Table 5). This was about 4 times higher than yield obtained from natural grassland (3.16 t ha^{-1} yr^{-1}). These results showed that it is possible to get more than 12 t ha^{-1} yr^{-1} biomass through established silvipasture on these degraded lands which otherwise was producing only 3 t ha^{-1} yr^{-1} through natural vegetation (Rai *et al.*, 1999).

Table 5 : Total biomass production under silvipastoral system and natural grassland at 8 years rotation

Components	Production (DM t ha^{-1} yr^{-1})
Grasses and legumes	4.55
Tree leaves and pods at 50% pruning	0.51
Fuel wood at 50% pruning	1.66
Timber at 8 years harvesting	2.39
Fuel wood at 8 years harvesting	3.18
Leaf fodder at 8 years harvesting	0.33
Total production	12.62
Natural grassland	3.16

Source: Rai *et al.*, 1999

About one-third of the lands in the Himalayan region are bouldery wastelands formed due to excessive erosion and deforestation on the hill slopes. These lands are virtually devoid of economic vegetation and can be rehabilitated by establishing silvipastoral systems. In the Shivalik foothills, silvipasture systems involving *Eulaliopsis binata, Saccharum munja* or *Vetivaria zizanoides* with *Acacia nilotica* are promising. Further, *Eucalyptus tereticornis* + *Eulaliopsis binata* has been identified as an economically - viable and eco-friendly system for rehabilitation of these lands. In the southern hilly region, a silvipastoral combination of *Acacia melanoxylon* and *Digitaria decumbens* on steep slopes was promising for soil conservation and meeting the fodder requirement. Similarly mixed plantation of *Eucalyptus globulus* and *Acacia mearnsii* - a system of a legume and non-legume ideal for soil and moisture conservation and also resulted in higher wood yield.

On the basis of research and potentials of silvipastoral systems under different agro-climatic regions of India potential species and their production potential are given in table 6.

Table 6 : Potentials of silvipasture system in various regions of India

Region	Components	Results
Himalayan Region I. Alpine	Only pastures with suitable grasses and legumes	*Agrostis* spp., *Chrysopogon fulvus, Dactylis glomerata, Poa pratensis* *Trifolium repens, T. pratense, Medicago falcata*
II. Temperate	Intensive pasture production, temperate horticultural plant combined with pasture legumes as ground floor, fodder forest legumes	5.7 t ha^{-1}, on an av. 8 to 10 kg $tree^{-1}$ dry matter
III. North-eastern region	Silvipasture, *i)Alnus nepalensis+ Stylosanthes guianensis Panicum maximum* *ii)Ficus hookerii + Thysanolaena maxima*	13.5 t ha^{-1}(DM) fodder yr. 12.3t ha^{-1} fodder yield
IV. Outer Himalaya	Silvipasture systems, *Dalbergia sissoo +Chrysopogon*	Suitable system for utilization of degraded lands for fuel,

	fulvus Acacia catechu + *Eulaliopsis binata* *Leucaena leucocephala* (high density) + Hybrid Napier (NB_5) *Bauhinia purpurea, Albizia lebbek* + *Eulaliopsis binata*	fodder and fibre 64-71 t ha^{-1} fuelwood 5.2-5.5 t ha^{-1} yr^{-1} fodder grass for short rotation of 4 -5 years Most suitable for gravelly riverbed lands of Doon Valley.
Shiwalik foothills	Silvipasture, Fodder production from wastelands and watershed management	Fodder production increased by 267%
Indo-Gangetic plains	Tree-grass combination Forage production on dry lands	Annual production of 8-10 t ha^{-1} (4.5-5 t ha^{-1} fuelwood) 7-10 t ha^{-1}yr^{-1} biomass production *Leucaena* + sorghum-safflower- Forage yield of 12.3 t ha^{-1}
Humid and sub-humid	Use of quick growing leguminous fodder plants in crop rotations, plantation crops with pastures, Farm forestry. Deenanath (*Pennisetum pedicellautm*) + *Stylosanthes* and Deenanath + *Leucaena* i)*Leucaena* (Hawaiian, Hybrid-28)+fodder grass. *Glyricidia/Sesbania/* *ii)Ficus hookerii* *Litsea+Thysalaena* */Andropogon nardus*	Best for pasture development in the eroded marginal and sub-marginal hilly area of Southern Bihar Suitable combination for plain area in West Bengal Suitable for hilly areas
Coastal area	Fodder trees (Subabul/ Agathis/ Desmenthes) + Congo signal + *Stylosanthes or Centrosema*	Total fodder yield of 22-29 t ha^{-1} could be obtained in coconut based fodder production systems.
Arid and semi-arid region	Grazing land and silvipasture and Hortipasture in arid and	Very good potential for silvipastoral enterprise

		integration of agri-silvipasture, Forage-cum-coppicing farming on the marginal and sub-marginal lands with intercropping of dryland cereals and legumes, *Acacia tortilis, Albizia lebbek, Prosopis cineraria* with *Chrysopogon fulvus, Cenchrus ciliaris, C. setigeres*

In Doon valley, silvipasture systems, such as *Dalbergia sissoo* + *Chrysopogon fulvus*; *Acacia catechu* + *Eulaliopsis binata*; *Eucalyptus* hybrid + *Chrysopogon fulvus*; and *Leucaena leucocephala* + Napier/*Panicum maximum* and *Albizia lebbek, Grewia optiva, Bauhinia purpurea* and *Leucaena leucocephala* with *Chrysopogon fulvus* and *Eulaliopsis binata* were found suitable for utilizing the degraded lands for productive as well as protective functions (Vishwanatham *et al.*, 1999; Samra *et al.*, 1999) (Table 7). Biomass production of *Eulaliopsis binata* (4.6 t ha^{-1}) was considerably higher than *Chrysopogon fulvus* (2.7 t ha^{-1}). There

Table 7 : Biomass production (kg ha^{-1}) of a silvipasture system after 14 years of establishment in degraded bouldery lands of Doon valley

Treatments	1993				Pooled (1980-1993)			
	Tree biomass		Grass	Total	Tree biomass		Grass	Total
	Leaf	Branch wood	Bio-mass		Leaf	Branch wood	Bio-mass	
TREE SPECIES								
Albizia lebbek	359	474	1078	1911	364	573	3592	4529
Grewia optiva	124	254	1994	2372	262	754	3461	4477
Bauhinia purpurea	395	459	2075	2929	362	629	3878	4869
Leucaena leucocephala	158	236	1338	1732	194	361	3655	4210
CD (0.05)	172	223	325	-	75	174	170	-
LOPPING INTENSITY								
50%	239	323	1572	2134	285	583	3567	4435
75%	279	389	1671	2339	306	575	3727	4608
CD (0.05)	NS	NS	NS	-	NS	NS	159	-
GRASS SPECIES								
Chrysopogon fulvus	302	423	1068	1793	371	723	2687	3781
Eulaliopsis binata	216	289	2174	2679	219	436	4607	5262
CD (0.05)	65	63	320	-	25	53	166	-

was a marked decrease in soil pH and increase in organic carbon, available P and available K in the surface layer due to the growing of the tree species of *Albizia lebbek* and *Leucaena leucocephala.*

Total biomass production

Green biomass production under two canopy management practices along with hybrid napier grass was compared with control in a silvipastoral system for different components and system as a whole (Table 8). The maximum green grass fodder (260 q ha^{-1}) from the Napier was obtained under pollarding management followed by grass alone without tree (control) while tree canopy of lopping has lower yield may be due to less infiltration of light. Whereas, pollarding canopy showed compatible association with Napier grass. The total biomass yield from trees was also recorded higher with tree pollarding alone (89 q ha^{-1}) without grass association followed by pollarding + grass (36 q ha^{-1}), which was about 53 per cent lower. The decline in the tree biomass yield is because of the effect of grass association (Napier). However, the total biomass from the silvipastoral system as a whole was highest with pollarding+grass (296 q ha^{-1}), followed by lopping+grass (229 q ha^{-1}). It was concluded that the pollarding+grass (management) was the best technique to obtain grass biomass from degraded lands following by lopping+grass. Secondly tree pollarding alone was the best technique for biomass production as compared with the canopy management practice with grass combinations.

Table 8 : Green biomass production (q ha^{-1}) under two canopy management practices from Hybrid Napier-*Grewia optiva* based silvipastoral system on degraded bouldery riverbed lands

Treatments	Green grass fodder	Total tree biomass	Total biomass from SPS
Lopping + Napier grass	201	28	229
Pollarding + Napier grass	260	36	296
Lopping only	-	30	30
Pollarding only	-	89	89
Napier grass only	221	-	221

Conservation of Forages

In order to sustain animal production, it is essential that the optimum feeding should be maintained throughout the year. However, there are 2 to 3 months of lean period (October to December/January and April to July) when the fodder availability to the animal is at its low. In the summer months it becomes more difficult to meet the maintenance requirements of the animals. It is, therefore, essential to conserve surplus forages at appropriate stage of maturity to feed the animals adequately during lean period for optimum animal production. The conservation of forages could be done in various ways especially in the form of silage from cultivated fodders (legumes and cereals) and also pasture grasses. Forages could also be conserved in the form of hay when dried to about 80-85 per cent DM which will preserve more of its nutrients including carotene.

Conclusion

A dynamic approach to tackle the problem of fodder and feed scarcity must consider how best to utilize the degraded areas both within and outside the forests, for forage production. Obviously, such areas have poor soil fertility and often suffer from moisture deficit. In such situations when one would think of forage promotion, he has to keep in mind a composition of forage plant types complementary to each other and not competitive in growth pattern and yield potentials. Thus greater emphasis needs to be placed on the leguminous component through the use of Rhizobial cultures.

References

Anonymous (2000). Agricultural Statistics at a Glance. Directorate of Economics and Statistics, Dept. of Agric. and Cooperation, Min. of Agriculture, Govt. of India, New Delhi.

Arora, Y.K. and M.K. Vishwantham (1995). Alternate land use system for marginal and bouldery riverbed lands. In: (Eds. G. Sastry *et al.*,). *Torrent Menace: Challenges and Opportunities*. CSWCRTI, Dehradun, p. 156-163.

Balocchi, O.A. and C.J.C. Phillips (1997). Grazing and fertilizer management for establishment of *Lotus uliginosus* and *Trifolium subterraneum* under *Pinus radiata* in Southern Chile. *Agroforestry Systems*, 37:1-14.

Deb-Roy, R. (1994). Common agroforestry systems and their management for optimizing production. In: Singh, P. Pathak, P.S. and Roy, M.M. (Eds.) *Agroforestry Systems for Degraded Lands*. Oxford and IBH Publishing Co. Pvt. Ltd., New Delhi, Vol. I, pp. 379-387.

Dhyani, S.K., K.S. Dadhwal and V.S. Katiyar, (1988). Biological measures for rehabilitation of abandoned mine area. *Indian Farming,* 38(7):15-18.16(1).

ICFRE (1993). *Annual Research Report for 1991-92.* Indian Council of Forestry Research and Education, Dehra Dun, India.

Jeschke, W.D. and W. Hartung (2000). Root-shoot interactions in mineral nutrition. *Plant and Soil,* 226: 57-69.

Jones, M., F.L. Sindair and V.L.Grime (1998). Effects of tree species and crown pruning or root length and soil water content in semi-arid agroforestry, *Plant and Soil,* 201:197-207.

Mathew, T., M.Kumar, B.V.K. Suresh and Maheshwaran K. Verma (1992). Comparative performance of four multipurpose trees associated with four grass species in the humid regions of Southern India. *Agroforestry Systems,* 17:205-218.

Nair, P.K.R. (1983). Soil productivity aspects of Agroforestry, ICRAF, Nairobi, Kenya, 85 pp.

Pathak, P.S. and R. Deb Roy (1987). Silvipastoral system for forage production on wastelands. *Indian Farming,* 36(8): 10-11.

Rai, P., K.R. Solanki and G.R. Rao (1999). Silvipasture research in India–a review. *Indian J. Agroforestry,* 1(2): 107-120.

Raison, R.J., B.J. Myers and M.L.Benson (1992). Dynamics of *Pinus radiata* foliage in relation to water and nitrogen stress: I. Needle production and properties. *For. Eco. & Manag.,* 52:139-158.

Samra, J.S., M.K. Vishwantham and A.R. Sharma (1999). Biomass production of trees and grasses in a silvipasture system on marginal lands of Doon Valley of north-west India. 2. Performance of grass species. *Agroforestry Systems,* 46:197-212.

Sehgal, J.L. and I.P. Abrol (1994) Soil Degradation in India – Status and Impact. Oxford and IBH Publishing Co. Pvt. Ltd., New Delhi, India.

Shankarnarayan, K.A. and K.C. Singh (1990). Rehabilitation of degraded pasture land in the arid regions of Rajasthan. In: Abrol, I.P. and Dhruvnarayana, V.V. (Ed.) Technologies for Wasteland Development, ICAR, New Delhi. p. 21-31.

Singh, B.P., S.K. Dhyani and R.N. Prasad (1994). *Traditional agroforestry systems and their soil productivity on degraded alfisols/ultisols in hilly terrain.* In: *Agroforestry systems for Degraded Lands* (Eds. P. Singh, P.S. Pathak & M.M. Roy) pp. 205-214. Oxford & IBH Pub. Co. Pvt. Ltd., New Delhi.

Singh, K.A., B.P.S.Yadav and S.N.Goswani, (1996). Forage resource management for sustained livestock productivity in Himalayan Agro-ecosystems. *Research Bulletin No.39,* ICAR Research Complex for NEH Region, Barapani, and pp. 94.

Singh, P. (1986). Forage production in India, RMSI, IGFRI, Jhansi, pp. 213.

Singh, Shamsher (2005). Management of degraded lands for sustainable agricultural production - An overview. In: Watershed Management for Sustainable Production, Livelihood and Environmental Security - Issues and Options. Lead Papers presented at the *National conference on Watershed Management*, May 19-21, 2005, Soil Conservation Society of India and G.B. Pant University of Agriculture and Technology. p. 30-41.

Upadhyay, V.S. and P.S. Pathak (1994). Silvipastoral system evaluation for secondary production in Central India. In: Singh, P., P.S. Pathak and M.M. Roy (Eds.) - *Agroforestry Systems for Degraded Lands*. Vol. II pp. 780-787. Oxford and IBH Publishing Co. Pvt. Ltd., New Delhi.

Vishwanatham, M.K., J.S. Samra and A.R. Sharma (1999). Biomass production of trees and grasses in a silvipasture system on marginal lands of Doon Valley of North West India. I. Performance of tree species. *Agroforestry Systems, 46*: 181-196.

Vishwanatham, M.K., J.S. Samra and A.R. Sharma (1998). Biomass production of silvi-pasture systems on gravelly lands of Doon Valley. *Bulletin No.T-39/ D-26*, CSWCRTI, Dehra Dun, p. 126.

Walker, J. J.A.Robertson, L.K. Penridge and P.J.H. Sharpe (1986). The influence of an *Acacia karoo* tree on grass productivity in its vicinity. *J. Grassland Soc. Southern Africa*, 4(3):83-48.

Wilson, J.R. (1996). Shade stimulated growth and nitrogen uptake by pasture grasses in a sub-tropical environment. *Australian J. of Agric. Res.*, 47 (7): 1075-1093.

Wolters, G.L., A. Maetris and H.A.Pearson (1982). Forage response to over storey reduction on loblolly-shortleaf-pine hardwood forest range. *J. of Range Management*, 35(4): 443-446.

Subject Index

c

d

e

f

g

h

i

j

k

l

m

n

o

p

r

FSC
www.fsc.org
MIX
Papier | Fördert
gute Waldnutzung
FSC® C083411

Zeitfracht Medien GmbH
Ferdinand-Jühlke-Straße 7
99095 Erfurt, Deutschland
produktsicherheit@kolibri360.de